NO PLACE
TO
HIDE

Recent Titles in Contributions in Sociology
Series Editor: Don Martindale

NO PLACE
TO
HIDE

CRISIS AND FUTURE
OF AMERICAN
HABITATS

Manuel Marti Jr.

Contributions in Sociology, Number 50

Greenwood Press
WESTPORT, CONNECTICUT
LONDON, ENGLAND

307.0973
MARTI

Library of Congress Cataloging in Publication Data

Marti, Manuel.
  No place to hide.

  (Contributions in sociology, ISSN 0084-9278 ; no. 50)
  Bibliography: p.
  Includes index.
  1. Human settlements—United States. 2. United
States—Social conditions—1980–   . 3. Twenty-first
century—Forecasts. I. Title. II. Series.
HT65.M37   1984   307′.0973   83-22762
ISBN 0-313-24271-2 (lib. bdg.)

Library of Congress Catalog Card Number: 83-22762
ISBN: 0-313-24271-2
ISSN: 0084-9278

First published in 1984

Greenwood Press
A division of Congressional Information Service, Inc.
88 Post Road West
Westport, Connecticut 06881

Printed in the United States of America

10 9 8 7 6 5 4 3 2 1

To my parents Olga and Manuel.
For the constant evidence of
wisdom, love and understanding.
I love you beyond words.

Contents

PART TWO   PROSPECTS

Figures

Tables

Acknowledgments

Some of the contents of this book have improved because of the review made by two colleagues and very special friends: Bennett and Kathy Lord.

This work is also in debt to the writings of Glenn H. Beyer, Daniel J. Boorstin, Edward T. Hall, Roy E. Mason, and Kirkpatrick Sale, who have provided extensive material from which to obtain facts and develop new ideas. None of these writers, however, may subscribe to my conclusions, some of which they may find unacceptable.

The manuscript owes a great deal to Judith Nystrom whose editorial and critical comments improved it vastly in both form *and* content.

And, finally, I wish to thank my wife Patricia, whose contribution to this book could only be properly recognized by sharing its authorship.

Introduction

In years to come the 20th century will probably be recognized as one of the most drastic transitional periods in the history of man. The generations of this century have witnessed developments of outstanding social and technoscientific impact: for the first time man ventured into the skies, the oceans and, finally, into outer space. Today's frontiers lie beyond the horizon. For the first time the surface of the Earth has been fully explored and chartered, its ecological conditions recognized, and man has become aware that he cannot take himself or his planet for granted.

Most modern societies have begun to integrate the effects of these transitional conditions within themselves. In all human groupings there are factors that cause and accelerate change (population, technology, the environment, resources, etc.), and the relative influence of these issues upon societal structures can be overwhelming. Today, even the most conservative cultures have had to give way, in one way or another, to the profound changes affecting the world.

Professor Gerard K. O'Neill, in his opening remarks to the 1982 American Institute of Architects (AIA) Convention in Honolulu, Hawaii, listed five major drivers of modern technological change (computers, automation, space colonies, energy sources, and transportation) and outlined the basic principles that development in any one of these areas should obey in order to cause change: improved performance and efficiency, and reduced environmental impact.[1]

Unfortunately, that analysis of the process of change is oversimplified, because while many of the forces behind technological development *have* operated following those basic principles, social change does not always conform to such simple formats. Improved performance, for example, only addresses preexisting technological conditions, and efficiency or reduced environmental impact are factors which have only become issues within the last decade or so. In context, technological progress and its sociological implications are a much more complex matter.

Transition implies change and change implies new rules. Because of this, man has often viewed the process of change as a threat. Traditionally, societies behave better in times when the interaction among their development factors (population, resources, technology, etc.) is balanced. Man has always felt more secure within stable cultural frameworks, and no matter how fast technoscientific change may have taken place, human nature has consistently resisted sudden transformations. In fact, many potential obstacles to future progress lie precisely in those areas of human endeavor which must deal with the interaction of man and technology.

The human shelter is just such an area. Habitats bridge a gap between technology and human nature. It is precisely in his built environments that man has been forced to find a compromise between his idea of himself and the realities of his existence, a balance between his technological flight and the ballast of his basic nature.

It is precisely because of this that throughout history the development of human shelters has had a moderate rate of change when compared to most other areas of human endeavor. In context, the *physical* differences between old habitats and new habitats of similar functions are relatively minimal. However, from the proper perspective their *conceptual* differences are quite noticeable. Despite minimal changes in man's basic nature, the social transformations of the last century have forced increasingly changing life-style patterns. And nothing illustrates the widening discrepancies between human nature and modern life-styles more clearly than the home of man.

Because of the multiplicity of their functions and symbolisms, human habitats are a complex area of study since they serve and represent not only man as an individual, but also man within society. And if contextually there have been relatively few significant variations in individual human habitation practices, socially the changes have been staggering.

In the beginning, the human shelter was essentially a refuge from an unknown, uncontrollable and sometimes hostile world. As mankind evolved, man started understanding, accepting and even taking advantage of the world's natural components more and more, and thus the concept of shelter began to change from temporary refuge to permanent dwellings that eventually expanded to large settlements. As scientific knowledge developed, shelters began to incorporate a greater variety of technological features, until eventually they became almost completely subordinated to the dictates of technology. Man had come to control his built environments and would continue to do so as long as he remained in control of his own technoscientific future.

However, throughout these changes the essential function of shelters remained the same: to provide a retreat, a place where the agricultural, industrial or technological human being could come to grips with his own self and his relation to his surroundings; the place where society and the individual could finally be reconciled.

More than a century ago Alfred Russell Wallace—the English naturalist who, concurrently with Charles Darwin, proposed the theory of the origin of the species through natural selection—noted that while animals evolved by adapting to the environment, man evolves by adapting the environment to himself. In this respect, recent arguments maintain that man seems to be moving into a rapidly changing environment which is the result of his own fast changing intellectual modes.[2] Contextually, there is an inherent danger in many of the current transitional forces affecting habitats, as more conditions that erode the basic role of shelters are forced within their envelopes, while the dwellings themselves become mere functions of rapidly changing life-styles and social prototypes that ignore some of the most essential aspects of the human condition.

The increased integration of technoscientific elements and conditions within edifices in recent decades has caused a relative dehumanization of man's built environments. As a result it is entirely possible that in the future, there may be no place left for man to retreat to, either by himself or with those close to him, because his shelter will have been made, penetrated and conditioned by the very forces he would seek refuge from.

Because the study of man's built environments must account for all facets of human experience, the following pages do not evaluate rapid technological or cultural change in terms of good or evil. This work does not advocate positive human evolution by sacrificing technoscientific advances, nor does it accept the notion that man must merely serve as a backdrop to progress. There are two undeniable realities in man—intellect and emotion—and both must always be recognized.

In years to come, this era may well be labeled one of the most significant transition periods in the history of man, and with reason; modern society is overwhelmed by the progress achieved in almost every area of human endeavor during the short course of three generations, and by the multitude of new horizons still waiting to be explored—without *and* within.

And so, more than simply a period of transition, this era could also be considered an essential turning point in the history of man, a time of challenges and accomplishments when a balance between scientific development, social change and the human condition was finally achieved.

NO PLACE
TO
HIDE

PART 1
ANALYSES

1
The Order of Disorder

The success of human shelters depends on the compatibility of their occupants' values with the characteristics of their design and construction. Whenever the importance of this concordance has been underplayed, problems have arisen, forcing revisions and readjustments. Whenever this basic principle, which links mankind to its built environments, has been violated, the final result has been failure.

Traditionally, the most important factors in the development of habitats have been the social characteristics and attributes of their dwellers. Aside from natural conditions, it is here that one finds the first clear traces of purpose and intent, and the complex gamut of behavioral patterns, religious beliefs, scientific knowledge, creativity and ingenuity, which have consistently linked human beings to their surroundings and social features.

Buildings will inevitably be found at the core of any study which attempts to understand a civilization. And with reason: it is here that one will find the clearest and most precise illustrations of the historical frameworks within which societies existed.

The basic qualities of buildings—function, form and construction characteristics—serve to unravel more than particular functions in time and space. When looked upon in context, edifices become the most reliable traces of mankind on this planet. As expression of man's goal of transcendence, they become testament.

The starting point for an analysis of the conditions from which this nation's shelters will evolve must be a comprehensive study of its social fabric. In this respect, the United States is a rewarding object of study since its cultural structure is markedly influenced by fads. The fabric of the North American society is embroidered with a complex myriad of interwoven contradictions that continuously redefine changing issues and trends by leaping to its surface or by forever sinking in indifference and oblivion. Strangely enough, because of these incessant cycles, the struggle for cultural survival has become one of America's most paradoxical social challenges.

Because of the complexity of this nation's collective structure, American shelters have a multitude of facets, house a vast collection of subjects and specialized conditions and are continuously affected by an endless list of interactive and conflicting issues.

A clear illustration of this last point, for example, was observed in the results of the Cambridge Survey Research polls taken during the 1972 presidential campaign. In a sampling of small rural towns in Nebraska, New Hampshire, Oregon and Wisconsin, the issue of "crime in the streets" surfaced as one of the major concerns of people who were asked to define areas of significant malaise within their specific communities. This was surprising indeed, since many of the sample communities had very few streets to begin with, and their occasional "street crime" over the previous four or five years had been limited to minor teenage vandalism of public property.[1]

The rural perception of crime in the streets, derived from the public media coverage of a problem affecting American urban nuclei, defined a *true* concern of citizens who had no real reason to worry about crime in the streets at all. And so, in striking contrast with previous decades, rural homes are being locked, their doors and windows carefully secured, and their curtains and shades drawn as the modern communication networks succeed in transmitting to rural areas an undue share of urban fears, isolationism and anxiety.

The increased sensationalism of the public media has culminated in a standardized overmagnification of *all* matters presented to the public eye. But consequently, individuals in the American society have developed an intricate self-defense mechanism of social indifference and isolationism to ward off this barrage of overdeveloped news. There is no other escape from the modern communication systems; all the protective tools of society (from laws to the enclosure of shelters) have failed to provide individual households with privacy. Americans have become subject to the most sophisticated public information system in the world and their habitats reflect this condition in a multitude of ways. Consider this: Never, at any point in history, have human beings seen the privacy of their homes usurped to a greater extent than it is by television, radio, telephone, magazines, phonographs or newspapers today. The modern public information media has gone beyond improving the social interaction among human beings; it has become the *only* social link among people who cannot escape the web of its communication networks.

However, not all critical issues affecting American habitats are consequences of uncontrolled communication sprawl. In the complex sociopolitical system which surrounds buildings and urban nuclei, improper measures adopted by public or private policies have often ended up aggravating the same problems they are designed to resolve. Take for example the last three decades of automotive expansion during which, to alleviate traffic congestion, billions of dollars were spent in the construction of new expressways and speedways by all major American cities. Yet, each time new expressways were built, more and more people found it easier to switch from mass transit systems to private vehicles. Thus the number of vehicles circulating on the expressway systems increased steadily creating new traffic congestions and, eventually, defeating the intent behind the construction of larger speedways. In spite of the fact that it has been this country's most treasured goal, America has never fully understood the implications and corollaries of growth.

Special interest groups have also been instrumental in developing policies of questionable effectiveness. Consider what happened recently to the housing market and the property tax structure in California. Due to extensive publicly-supported no-growth policies, initiated mostly by neighborhood associations and effectively implemented throughout most of the state, the average sale price for homes in Orange County jumped approximately 30 percent between 1975 and 1976. And in 1977, the cities of San Francisco and Los Angeles led the country in housing prices.[2] Because of these artificially inflated new costs, valuations used as bases for property tax assessments throughout California rose to figures well beyond those which the same property owners who had so wholeheartedly supported no-growth policies were willing to tolerate. The result was Proposition 13: a cut in property tax collection in excess of 50 percent. Neighborhoods, which had been made artificially exclusive by driving up costs through no-growth policies designed to prevent new housing developments and the opening up of new lands for residential subdivisions, were refusing to pay the price of their action. Yet the only lesson this nation supposedly learned from Proposition 13 was that government spending should be reduced. It is a valid lesson, but somewhere along the line a second lesson of yet greater importance was lost: driving up the value of homes artificially by making neighborhoods exclusive, through inaccessibility and unaffordability, is a decision that must be paid for.

At the heart of these urban/suburban struggles lie deep-rooted social differences. One basic problem is simply that urban conglomerates do not fit into precise molds or statistical formats. Even if most houses in new developments look alike, within a matter of months individual rearrangements will disguise most similarities. The "personalization" of space is an inherent human characteristic. However, although society is made of quite different individuals, modern habitats are essentially aimed at average models. And this essential disparity between human nature and modern spatial standardization is another characteristic of American built environments that appears to be reaching a critical stage.

Likewise, national urban patterns are engulfed in a system which is slowly being immobilized by an increased fragmentation of components, regulatory impositions and conflicting value systems. Many urban renewal projects fall into this category since, despite what has been preached, the essential philosophy behind most urban renewal appears to be that one *should* judge a book by its cover.

ECONOMIC FACTORS

In the last decade the new economic dilemmas brought about by increased government spending, soaring interest rates and declining industrial outputs have seriously damaged real estate transactions and made new construction ventures extremely risky.

Near the middle 1980's, the nation appears to be moving closer to the end of the American dream of home ownership, with an approximate 157 percent increase in home prices in the last ten years alone.

Inflation, monetary policies, land use restrictions and other regulatory constraints have finally taken their toll on the housing industry. It seems plausible that the time when young Americans are priced out of a home of their own may be near.

Assuming 1980 as a comparative reference, the median home price of about $63,000 represented an 18 percent increase from 1979 alone. For the first time buyer, however, the figures were worse: up approximately 31 percent for new housing, while their median income rose only 14 percent; and their average monthly payment (once soaring interest rates affected the cost of money) jumped 33 percent, which actually represented a net increase of almost 7 percent (from 29 percent in 1979 to 35.6 percent in 1980) as a percentage of their income. Down payments also climbed, approximately 3 percent in one year.[3] This country could be approaching a serious housing supply crisis.

The effects of these conditions added to the radical decline of real estate ventures of the last decades have been disastrous for the development patterns of cities and habitats. Take, for example, the "immobility predicament" faced by most homeowners. Modern society has emphasized and rewarded for decades the need for physical mobility as a competitive advantage in job markets, thus making housing construction one of the best indexes of the overall state of the economy. Parallel to the economic growth of the postwar years, for example, the number of real estate transactions increased significantly throughout the fifties.

It was during the following decade, however, that the need for mobility began to affect the North American work force. The sale of mobile homes and the proliferation of trailer courts soared to dramatic levels, rising from under 9,000 units in 1960 to over 38,000 units in 1970, a fourfold increase in ten years.

The phenomena presented by the gradual change from permanent to mobile shelters in this country has been the inadequacy of American housing with regard to the needs of a changing society. Traditionally, houses have been a symbol of settlement, permanence and immobility. Other than purchasing a form of shelter which partially satisfies both ends of the dilemma (such as trailers, vans, motor homes, etc.), the need for social mobility leaves Americans with buying and selling as the only means of overcoming this intrinsic housing permanence in a nation where ease of relocation is essential.

High mortgage rates or economic declines further complicate and impair the few available solutions to this inherent conflict between American habitats and their unstable dwellers. It has been observed that 50 percent of the population changes residence every 5 years and some have generalized that the average person will move approximately eight or nine times in a lifetime.[4] As we will see later on in this text, lack of permanence is one issue which has had (and will have) an enormous effect on the development of this nation's built environments.

Another significant consequence of increased financing costs is that it has brought about a surge of "do-it-yourself" home refurbishments and consequently, new construction products and techniques, as homeowners try to remodel their dwellings rather than follow the common behavior of moving to a new house.

In the area of urbanization, present speculations about mobility and change appear to contradict one another. On the one hand, we have those who forecast the death of the large urban nucleus, while on the other there are those who announce a revitalization of the inner city.

Both sides have valid points of contention. Those who claim that large cities are beginning to experience a slow death base their arguments on statistics of migratory population trends which show an increased tendency towards people moving *out* of the city rather than *into* it. According to the Rand Corporation, in the 1970–1975 period, for every 100 Americans that moved into metropolitan areas, 131 moved out. And towns of 5,000 people or fewer are growing at faster rates than other urban nuclei. Also, the general population shifts between 1970 and 1979 show regions with high population density (such as the Northeast) experiencing higher population outflows than other sectors of the nation.[5]

In addition, smaller cities show a tendency of climbing to certain levels of population and services, but after reaching this peak their ratio of growth decreases or slows down significantly. The reasons for these trends are overcrowding, high prices, taxation and the inefficient services that characterize large cities. These factors have also contributed to the growing trend towards decentralization which is one of the aspects of urban development most widely commented on by forecasters.

In opposition to those who foresee continued urban decay are those who predict a golden era for cities, basing their argument on such developments as increased transportation costs, effective measures to combat crime (such as the closing of streets in residential areas), successful self-contained communities within city limits (like Marina Towers in Chicago) and large urban revitalization projects which seem to be performing satisfactorily, such as the Renaissance Center in Detroit, Michigan. Decay of the suburban areas adjacent to large cities is rising and has also been cited in support of the argument for inner city redevelopment. However, this broad generalization is not applicable to all cities.

SOCIAL FACTORS

Everything that impacts the structure of society becomes a key determinant in the future development of its habitats. During this decade the postwar "baby boom" will reach the age at which most couples become homeowners, causing the need for housing to exceed that of previous years. The question is: *which kind* of housing?

To understand the vital significance of that question one must realize that the "typical" family of working father, nonworking mother and two children constitutes a very small percentage of all American households in this country and that if to the social mosaic created by the growing number of household types in America one adds the soaring costs of single family dwellings, the price of land, insurance and taxation costs, etc., it might be reasonably speculated that the prefabricated and mobile home industries, which today provide over 25 percent of all new housing in this nation, could very well end up providing for as much as 50 percent of all American dwelling needs within a few decades.

Innovations in habitats, on the other hand, can only take place gradually. If one assumes the average age of an American house to be 60 years, it is unquestionable that well over 70 percent of the housing this country will be dwelling in by the year 2000 has *already been built.*

Social characteristics are essential determinants of two of the most significant factors that structure the future of habitats: style definition and spatial allowances. Human built environments are direct reflections of the culture that builds them. To illustrate this point and its implications briefly, consider the changing role of the bathroom within the last 80 years.

Bathrooms are relatively recent spatial requirements closely linked to the late 19th century availability of hot and cold running water and sanitary sewer systems in cities. Their primary functions dealt with personal hygiene, privacy and human interaction with specialized artifacts whose purpose and ergonomics were aimed exclusively at the cleanliness and waste disposal of the human body.

Now, compare the changes in the role presently played by bathrooms within the functional complex of the dwellings: they are essentially still areas of hygiene, privacy and extensive human interaction with specialized artifacts, except that now their purpose and ergonomics are also intended to provide for the user's physical and psychological comfort, privacy and relaxation, specialized treatments, and a gamut of satellite services ranging from the verification of weight, facial makeup and aging treatments to the performance of sexual functions. Because of this functional pluralism, modern bathrooms are filled with a collection of automatic controls, specialized equipment and related gadgetries, which permit the disposal of human residues to be performed in an impeccably prophylactic manner.

The essential system at the core of the bathroom, plumbing, has not changed in principle since the days of the Roman Empire. Obviously the functional differences and uses are a product of something else. A product of what? The answer to that question is also found in the human shelter.

TECHNOSCIENTIFIC FACTORS

In 1951, when computers first began to find their way into businesses and services, it was predicted that within 30 years we would probably have a "paperless society" where computers and automation would have practically "taken over" offices. Not only did that forecast not come true, but in that same period paper consumption in the office rose by more than 30 percent. Strangely enough, undaunted by this previous failure, and after a short period of relative silence, prognosticators have revised the forecast: the span is now 30 to 50 years.

Unfortunately for most scientific and technological predictions, the actual capabilities of science and technology do not always keep up with boundaries of human imagination.

Traditional building materials—whether renewable or nonrenewable—rank among the most prevalent resources on Earth. Here, aside from consideration of structural and physical adequacy, the process of selection has been greatly influenced by availability. Despite this, several recently developed products (products which are not obtained by direct mining or simple manufacturing processes such as stone, brick, or wood) are gaining popularity as their widespread use extends to areas of construction which, up to a few decades ago, were the exclusive domain of traditional construction materials and systems. The widespread use of plastics, fiberglass, stainless steel and concrete admixtures, among others, has brought significant advantages to the development of American habitats. Consequently, the extensive use of these materials has created an increased dependence on a new generation of factory-produced and industry-controlled products which are subject to the same basic economic laws of supply and demand, inflation, recession, labor strikes, regulatory controls and quality fluctuations as any other man-made item. So, although a basic raw material may be abundant, the availability, cost and quality of a specific finished product derived from it is continuously going through the same variable cycles as any other item in an industrial society because of the increasingly complex industrial processes destined to transform it. How strongly modern technology has changed the basic concept of building shelters from surrounding or readily available materials is illustrated by the fact that today it is practically impossible to build *any* kind of structure without the aid of specialized tools, equipment and products. Unquestionably, the one thing most modern technologies can be sure to produce are more technologies.

Presently there are limitless uses for inflatable structures, modular construction, laser beams, underwater construction, computer applications, photovoltaics, passive solar and earth-sheltered structures, and, in general, a myriad of technical possibilities ranging from the simple use of drafting machines to outer space industries. But there are also such essential issues as ecological imbalances, nonrenewable resources, acid rains, atmospheric "greenhouse effects," beaches that erode and forests that disappear.

So, prior to determining what courses of action might be the perilous routes or "yellow brick roads" to progress, one must first understand yesterday's paths and trace an accurate map to today. Only then can the true signposts to tomorrow be properly identified.

2
Facts on File

To venture into a study as complex as human shelters, one must have a clear understanding of the basic factors which define the development of habitats throughout this nation. Factors like energy, population, resources availability, the environment and various sociopolitical issues are inherently linked to the future of buildings and communities. The following is a brief summary of these factors.

ENERGY

Although the construction industry itself is not a major consumer of energy, buildings constitute one of the most serious areas of concern regarding energy waste. Approximately 40 percent of the energy consumed by the United States is used to provide homes, commercial structures and factories with artificial climate, light and hot water. The residential sector alone uses 20 percent of all the energy employed nationally. For decades now, buildings have been designed and built with almost complete disregard for energy efficiency. During the 60's, the energy efficiency of new buildings and homes was neglected to such an extent that between 1960 and 1965 office buildings in New York City were using twice as much energy per square foot as those built between 1945 and 1950 due to the introduction of fixed glass panels in lieu of operable windows and electrical and mechanical equipment for lighting, heating, ventilating and air conditioning systems.[1] Similar trends were followed in private dwellings where convenience, fashion or simple disregard for energy consumption led to the proliferation of energy-inefficient household appliances.

Today everyone is extremely conscious of energy-efficient building design, but just slightly over 10 years ago the newly rediscovered *solar* design was completely ignored in this country. Some may proudly announce that modern shelters should be designed to be cool in the summer and warm in the winter, making it necessary for houses that look towards the south to have overhangs or porticos so as to permit the winter sun to shine into the rooms, while in summer, the sun, passing high above the roofs, throws those same rooms into shadow. And thus, they will only be proudly repeating the very same words that Socrates used in or about 400 B.C. while discussing dwelling arrangements.

Unquestionably the oil crisis of the early 70's brought about a real surge in energy consciousness which has reached beyond the concern for petroleum resources and forced a close look at the entire energy spectrum.

In this regard, while the proposition that coal be used as a transitional energy source seems destined to perish in a sea of environmental regulations, its long-term prospects look good as a complement to this nation's transitional energy needs. Nuclear power, on the other hand, has fallen prey to social unpopularity. The present challenge appears to lie in the development of safe energy sources which do not depend on nonrenewable resources. Hence the importance attributed to solar energy. It is entirely possible that eventually solar technologies will provide America's badly needed unlimited energy source. Unfortunately, as of this writing, technologies that would permit solar energy sufficiency on a dwelling unit basis have not been proven feasible yet. This brings us to the acknowledgment of the one energy source involving no radioactive waste, no petrodollars or pollution, and limitless potential: conservation, a key variable in any effective equation attempting to define the immediate future of American habitats.

POPULATION

The population growth projections for the coming decades indicate that feeding, clothing and sheltering the future inhabitants of this planet will be indeed a serious problem.

In this light, health care and housing, with a minimum standard of quality, will be essential. It is noteworthy that the greatest achievements in health care have involved advances in hygiene and disease prevention facilitated by improvements in services such as plumbing, potable water, sewers and electricity, which today are an intrinsic part of the modern concept of "shelter."

The dangers of population explosion are not confined to a few underdeveloped nations. Eventually the United States may find itself dramatically affected by world population growth, even if it maintains its own present no-growth birth/death rate, through increased migratory influxes and worldwide socioeconomic crises.

To these conditions one must add a significant factor: population implosion—the further concentration of people in urban areas or specific regions—a trend which is likely to continue, making the improvement of shelters and their supporting services much more difficult throughout the world. Projections for the next two decades place urban populations in the threefold increase range worldwide. The more people and fewer shelters there are (since housing *does not* follow exponential growths of such magnitude), the more crime, loss of privacy, less sanitation and worse living conditions there will be.

The peak of global population explosion *may* have been reached already, but the world still must face the task of housing 2 billion more human beings during the next 25 years.

ENVIRONMENT

The most significant environmental issues, which will impact the development of habitats throughout this nation, are atmospheric pollutants, acid rain and hazardous materials.

In urban and suburban settings, atmospheric pollutants and high acidity levels are slowly corroding buildings and historic landmarks all over the world. Most of the monuments in Washington, D.C., are showing visible signs of damage, and it is forecast that in the not too distant future the inscriptions on the marble markers at the Arlington National Cemetery will be totally unreadable.

In addition to specific regional impurities, urban air has been found to contain sulphur dioxide, dust from furnace ashes, asbestos particles from automotive brake linings and building materials, mercury vapors from industrial operations and other organic compounds. Statistics *do* show that people living in polluted urban air experience more disease than people living in less polluted (usually rural) air.

Rain has always been slightly acidic, but in the past 25 years rainfall has become 40 times more so. The marble exterior of Chicago's Field Museum of Natural History has undergone such major decay that it is estimated that within 20 years its exterior features will be completely unrecognizable; and presently a newly developed multi-layer epoxy-based coating system, Versacor, is basing the bulk of its promotion on its presumed resistance to acid rain. Meanwhile, Lake Erie is slowly becoming an enormous underwater cesspool because of the dumping of municipal and industrial organic wastes and the runoff of agricultural fertilizers.

The microenvironment which surrounds the individual is also one area of concern. Modern life provides a gamut of facilities and comforts, but it has also introduced new materials and conditions into edifices which, in a very real sense, have reshaped human habitats.

Dust, trapped in houses and buildings with artificial climate and containing particles of toxic and allergenic materials, is gradually modifying the environment of shelters. House dust has been found to contain plant particles such as mold, cellulose and pollen; animal fragments such as mites, fur, hair, etc.; and man-made materials such as fiberglass, paints, cigarette smoke, fireplace soot, etc. Likewise, exposure to newly developed products like lacquers, varnishes, plastics, polishes, insecticides, paints, fuels, etc. can seriously affect anyone's health.[2]

The modern American shelter has also been found to have increased electrical charges in its air and to be extremely subject to electromagnetic waves and radiation, whether from its own internal equipment and appliances, such as radios, televisions, microwave ovens, stereos, dishwashers, telephone and alarm systems, or from external sources such as radio and television stations, high transmission lines, CB radios, radar installations and communication satellites.[3]

When urban air is overloaded with poisonous mercury vapors, yet mercury itself is a scarce raw material, certain environmental cycles deserve more attention. However, how can modern society eradicate collective habits or dependencies which have become an intrinsic part of itself without endangering its own structure? The answers to this and other similar problems will constitute formidable factors which will unquestionably shape the future evolution of human built environments.

NATURAL RESOURCES

The concern for fuel shortages has increased the popular awareness of energy conservation and efficiency, but it has also foreshadowed the shortages of several raw materials which could become critical in the future since, unlike oil, there are no known substitutes for resources as basic as water, or metals such as manganese, chromium and cobalt.

Iron and steel cannot be manufactured economically without manganese, and stainless steel cannot be made without chromium. Yet for these, as well as for many other raw materials, the United States now depends on imports to meet anywhere from 50 to 100 percent of its needs. The Federation of Materials Societies has recently issued a major policy statement warning that a national policy regarding the use and consumption of materials is vitally needed. To be specific, the statement notes that the United States now imports more than 90 percent of its total annual requirements of bauxite, chromium, cobalt, columbium, manganese and platinum. It also notes that present imports of tin and nickel equal 75 to 90 percent of the national needs, and imports of antimony, cadmium, tungsten and zinc account for 50 to 75 percent of the national consumption.[4]

With regard to building materials specifically, unless some careful recycling or control measures are adopted, several critical shortages can be anticipated before the turn of the century. World demand for gypsum through the year 2000 is estimated at approximately 2,540 million tons with less than 2,100 million tons of identified reserves. And barely meeting the anticipated requirements through the end of this century are the known reserves of zinc, lead, tin, copper and talc. Furthermore, the United States share of several of these essential construction materials represents only a fraction of total world reserves, while its consumption far exceeds the supply.[5] A shortage of gypsum (considering the widespread use of gypsum products in construction) should be a matter of grave concern in itself; but if to this one adds that the building industry in the United States relies almost exclusively on gypsum products for fireproofing, and that one of the possible alternatives, asbestos, is not only another material threatened by worldwide shortages but also an identified health hazard, the situation then becomes much more serious.

SOCIAL ISSUES

America's cultural ramifications have expanded to such an extent that its essential framework cannot be charted within any simple classification.

Yet, throughout this process of fluctuating sociopolitical, cultural and economic patterns, this country has maintained its ability to continuously renew itself by accepting change as an inherent social characteristic. How this process has affected the basic structure of many sectors of the North American society is clearly apparent in the transformation of its household structures, its sociopolitical interactions and its communication patterns.

Households

The spatial program of the modern single family dwelling has been severely affected by the enormous proliferation of different household units that has taken place in this nation since the Industrial Revolution. Perhaps the time has come to seriously question whether the "3 bedroom, 2 bath" detached house is still an adequate representative of the ideal North American home. The housing unit to be built within the next few decades will be aimed at a much broader family concept and will be defined by factors that are just beginning to surface.

Communications

The extent and complexity of communication networks in this country is another area which has reached surprisingly high levels of development and sophistication in just a few decades. Television, for example, which began with regular broadcasts just over 40 years ago (1941), grew worldwide from one million receivers in 1946 to approximately 50 million in 1959, with about half of those units located in the United States. By the mid–1970's this nation led the world with well over 110 million receivers.[6]

If to this explosive growth one adds professional broadcasting, newspaper, magazine and book publishing, telephone and telegraph, amateur and citizens band radio, computer systems and the United States mail, it becomes clear that the communication networks are a major driving force behind the spatial evolution of North American habitats.

The buildings erected in this country since the end of World War II have clearly reflected this communication revolution in the spatial allocations and engineering systems implemented within their envelopes. Modern edifices are not merely dwelling structures; they are transmitting and receiving centers as well. Any average building today has at least seven or eight different types of communication channels, of which two at least (mail and telephone) are capable of transmitting as well as receiving personalized information on a daily basis.

Houses and buildings have become theaters of the fantastic or peepholes to the outside world, but their most essential function of sheltering their dwellers from external intrusions has broken down since in modern shelters the intruders live within. If the question, "What is truly an American home?" were to be answered by someone who had thoughtfully analyzed and evaluated its traditional basis and essential characteristics, the response would have to be that whatever *is* truly a home today is quite different from what *used* to be a home years ago. Somewhere, somehow, the essential concept of "shelter" *has* changed.

GOVERNMENT

Countries grow and their societal structures become more varied, thus increasing the complexity of the governmental structures designed to rule them. Developed nations will inevitably have more complex regulatory structures than underdeveloped ones.

The United States has gone one step beyond: it is overdeveloped and overregulated, not unlike the mouse that crawled through the hole of the storage room and ate so much that it became a prisoner of its own appetite.

There is a price to be paid for this. Nowadays government regulates people's lives to such limits that there is not a single social activity left that has not been fully considered, classified, modified, tampered with and codified by a myriad of regulatory powers.

The United States government has increasingly expanded its influence in the evolution of shelters since early in this nation's history, but in the last few decades, its role has been decisive. During the Cold War, for instance, the American legislature's fear of the Communist bloc worldwide domination plans became the primary motive behind emergency government layouts which, by federal regulation, have been reminding the entire population of this nation since 1950 of the ever-present possibility of a nuclear attack, through evacuation drills, emergency sirens, radio station tests, fallout shelters, instruction posters and emergency government centers in counties all across the nation.

In a similar fashion, one finds the traditional characteristics of the single family dwelling modified to create self-sufficient fortresses, strategically located according to "national security maps" in areas of assumed safety from crime and radioactive fallout, and equipped with exposed plumbing for easy accessibility, emergency generators, wood stoves and survival appliances with optional fallout shelters, secret storage facilities and even stocks of survival rations and ammunition.

Policies of overinterference and excessive regulation extend to almost every corner of the American scene. Public housing is one of those areas where the regulatory burden has done the most damage. Almost every profession or trade presently involved in building construction agrees that the greatest problems faced by the industry have their origin in excessive or inadequate regulatory constraints, ranging from costly and inflationary requirements to conflicting and unclear laws.

In the last 70 years or so, the federal government has gained powers that far exceed those it was originally designed to have. Yet, as it grows, this power seems to become increasingly ineffective: despite cutbacks, the Office of Worker's Compensation in the Department of Labor is asked to look at approximately 40,000 claims per year, each of which takes an average of over 600 days; therefore, to give any disabled worker service on his/her claim within a year, this office would require a staff of approximately 94,000 investigators, which amounts to approximately *six times the total number of employees for the entire Department of Labor*.[7] Obviously, when the regulation that allowed for claims was designed, something very important was not taken into account: how to make it work.

In a similar fashion, many of the guidelines of the Occupational Safety and Health Act (OSHA) requirements regarding building construction are extremely difficult to enforce. Regulatory mandates, affecting building design and construction, however, do not cease to multiply.

In Sheldon, Iowa (population approximately 4,500), the mayoral office is expected to fill out yearly about 27 feet of government forms in quadruplicate, most of them concerning minority employment. In Sheldon, Iowa, there are no minorities.[8]

CROSSROADS

Because of the speculative nature of forecasting, the reliance on excessive quantifications may lead to inaccurate conclusions.

Disraeli said that there are three kinds of lies: plain lies, damned lies, and statistics. Often the presentation of a statistical result is manipulated to such an extent that its ultimate form obscures its true meaning. The statistical size of the average American family, for example, consists of 2.35 individuals. That could translate to a real estate developer trying to program a typical housing unit into a 2 bedroom, 1 bath dwelling. But if one considers that this theoretical 2.35 occupant figure is only an *average* which represents the combined results of many different family sizes the statistical data is, for the most part, useless and misleading. The distribution of household sizes *as percentages of the total population* shows a much different picture. Actually, 2 bedroom dwelling units end up satisfying fewer than 50 percent of the total American households.

In statistical and historical contexts, the quality of the North American built environment has improved significantly, but this is not enough. A brief glance at the basic issues outlined in the preceding pages cannot possibly lead anyone to the conclusion that *all* is well. The impact of unchecked modernism has affected American habitats dramatically; never before had the home of man been reckoned so impersonal or dangerous. When buildings are plagued with conditions hostile to human life, how can anyone justify their existence? The standards set by contemporary regulatory agencies claim to be the highest in the world, but for every effective measure there seems to appear a new challenge: in response to an endless list of fire and safety regulations, recently released figures show that arson accounts for about 45 percent of all losses (in excess of 50 million dollars per year) suffered by a given insurance company, and that motives including fraud, revenge, burglary or mere vandalism represent almost 80 percent of the causes behind arson claims, of which over 50 percent involve private dwellings.

What are human beings to do when the buildings, which are supposed to protect their properties, valuables and lives, become mere commodities which are subjected to this type of crime? Buckminster Fuller dreamed of portable and hygienic plastic bathroom units, but like many other scientists in love with their world of inventions, he forgot about the rapist who would then only need a cigarette to burn his way into his victim's refuge.

The limited effectiveness of most regulations dealing with the development of shelters gives rise to the controversial consideration of their ultimate need. Thus, the country finds itself facing the ironic dilemma of a nation born of the revolt against a large, overbearing and overtaxing foreign government, but now engulfed by an enormous overbearing and overtaxing government of its own.[9]

In sum, these general reviews of basic issues merely serve as a means to formulate a basic set of questions from which to develop more detailed analyses.

To fully understand how the spatial layouts of American shelters will develop as new sociopolitical and economic changes take place in the forthcoming decades, it is essential to comprehend the phases involved in the functional transformations of buildings in this nation throughout history. How habitats related to specific cultural and territorial conditions, how settlements evolved, and how social interactions, economic factors and technological achievements affected the formal and functional developments of the American built environments will be analyzed in the next three chapters.

3
Forms and Functions

This brief outline of the development of American shelters to date will be divided into three basic periods: colonialism (1600 to 1800), expansionism (1800 to 1900) and modernism (1900 to present), underlining within each period the patterns of distribution, fragmentation, growth, regionality and cultural characteristics of habitats and their interaction with social conditions and technological developments.

In earlier times, most buildings erected were initially dedicated to housing and, as communal growth took place, the structures developed for public use were assigned multiple functions not necessarily compatible with one another, such as religious or political gatherings, storage of materials, etc. Because of this, the development of housing is a better representative of the patterns followed by early shelters than that of any other building type.

COLONIALISM

The first recorded housing developments established in the United States date back to the early 1600's. Because the early colonists came from various countries and settled in different climatic regions, their first homes had a wide variety of particular characteristics from the start.

Yet there were strong communal similarities. For these early settlers the move to the new continent was permanent, and temporary shelters are only observed in the earliest dugouts and rudimentary structures constructed to provide refuge while more permanent dwellings were erected.

Another similar characteristic of these early communities was their attempt to duplicate the housing formats, spatial patterns and architectural styles of their homelands. The colonists also brought with them the tools, equipment, building systems and construction techniques of their mother countries. Not all settlers were capable homebuilders, but records indicate that, through a primary division of specialized labor, craftsmen, sawyers and other construction-related occupations could be easily identified from the start.[1]

Colonial houses were rectangular in shape and had low ceilings, so they were relatively easy to build, heat and maintain. The functions of cooking, dining, living and sleeping were carried out within the same confines. A large fireplace was used for cooking and heating and a loft, accessible by ladder, served as a sleeping place for children and as a food storage area.

As lean-to's were added, cooking and other activities were transferred there. When this happened, the one room cottages became two room houses where one room served for general living purposes and the other was reserved for receiving important guests and family gatherings, or sometimes it served as a guest bedroom.

This primary fragmentation of functions in the colonial home is a very significant characteristic. The more intimate and informal household activities were assigned to a specific room, the hall, while formal social interactions gained special recognition with the introduction of a new area—the parlor, undisputed ancestor of the modern American living room. Social evolution had begun to modify the utilitarian characteristics of colonial shelters. Not surprisingly, it was in this parlor that the family's best furniture was kept.

There was no demand for, nor would there have been any need to erect, monumental buildings in the colonies. Most of the public buildings of the 17th century had domestic characteristics and uses. Monumentality in governmental architecture did not appear in the colonies until the second half of the 18th century, when books, artwork and other forms of communication about European styles and buildings began to spread new spatial and stylistic influences.

As commerce and the fishing industry developed, the colonies prospered and new wealth found its architectural expression in the so-called Georgian style. In housing, this style was characterized by the introduction of several nonutilitarian features, such as temple porticos, entrance halls or "vestibules," and decorating windows evenly spaced on either side of doorways. Further plan fragmentation separated the kitchen, dining room and living room functions.

One of the more striking differences between the early colonial dwelling and this adaptation of the Georgian style lies in the changes observed in relationships within the family and between the family and the community. The early colonial dwelling provided common space for every member of the household and their guests; furnishings and even utensils were collectively shared. Individual privacy as defined today did not exist, much less the spatial separation of activities. A visitor entering these houses stepped directly into an area where any family function was readily perceptible. In marked contrast, a visitor entering a Georgian house was welcomed by an unheated hall, where doors isolated the different rooms in which the family carried out its functions.[2] Eventually, these complex spatial fragmentations separated not only visitors from family members, but family members from one another as well. Somehow, the social development and prosperity of the colonies seems to have favored this evolution towards European patterns of spatial fragmentation, since the layout of what was considered an adequate American house in the 18th century was much more fragmented than its counterpart of the 17th century.

Before 1750, colonial public structures were religious, educational or governmental; but by the third quarter, new building types began to multiply rapidly: hospitals, refuges, asylums, workhouses, prisons, communal structures, etc., most of them incorporating the colonial versions of the styles in vogue across the Atlantic. This transformation of building "areas" into building "types" followed a fragmentation pattern very similar to that of individual dwellings; and so, the spatial specialization extended from housing to public buildings and began transforming the shelter usage of families and communities alike.

Following the Revolution, the strong neoclassical influences of the Federal style changed many general design features in residential and public buildings, and temple-like features became basic architectural elements of the first truly "American" governmental structures, thus defining the monumental traits which have characterized public buildings to this date.

In general, the patterns of development of most colonial dwellings followed a similar course throughout the New World. In particular, however, cultural differences, climatic conditions and construction techniques gave way to distinct building characteristics. The Swedes, for example, were responsible for the development of a historic type of house: the log cabin, a three room cottage partitioned near the middle with another wall dividing one of the halves into two smaller rooms. The large room was used for living, cooking and eating, one small room used for sleeping and the other as an entry. The log cabin is one of the clearest examples to date of utilitarian shelter. Tree trunks were used in full shape or somewhat squared out of the rough; branches were cleaned off and the trunks evened in length, but not turned into man-shaped boards; they were laid upon one another as necessary and provided with notched ends, one letting into another in such a manner that the entire structure was erected without the use of nails or spikes.

In general, the houses of the early settlers were rough and simple structures, well adapted to the local conditions, but the materials for construction were scarce. Nails were costly and dwellers were known to set fire to their homes before leaving in order to recover them. Eventually, governments had to commit themselves to providing each family with an estimated number of nails per shelter to prevent the destruction of buildings as they were abandoned. Wood was the most widely used construction material and brick was sometimes used for foundations or for chimneys.

Lighting in colonial dwellings was provided by fireplaces, home-made candles or rush lights. Plumbing services were nonexistent, so human wastes along with garbage and other refuses were generally deposited in open pits dug in rear yards which were eventually back-filled.[3] Life in the early colonies was far from glamorous, but in many ways it possessed a very special character of its own: a certain blend of individuals and communities in harmonious coexistence with their surroundings and their natural habitat. Nevertheless, social change and economic expansion were not to be denied. At the turn of the century new transformations began to surface.

EXPANSIONISM

Throughout the 19th century, life in the United States followed two basic patterns of development: the continued economic growth and cultural evolution in settled regions, and the dynamic life of the frontier. The frontier expansion extended far into the deep South, where several plantation communal nuclei were developed around palatial homes, although the majority of the southern frontiersmen lived in one or two room cabins.

In the Midwest, a mixture of log cabins and frame cottages were the initial dwellings of the frontiersmen. Originally, the pioneer homes in these areas were variations of the colonial log cabin. On the average, the conversion from log cabin to frame house took approximately 20 years. Sturdier houses of brick, stone or wood with full brick basements (and cyclone refuges in the open country) emerged as small villages and towns began to sprout.

The public buildings of the small frontier towns that were built during the early expansion periods had utilitarian and simplistic appearances very similar to those of earlier colonial structures. Accordingly, inns, hotels, general stores, saloons, jails/law enforcement buildings, town halls, or commercial and service structures were made primarily out of wood, adobe, or a combination of the two (depending on their location), and developed adjacent to one another along the sides of a "main street" artery that served as the center of government, trade, commerce and social activities. Later on, as more established settlements were defined, public and commercial buildings became more permanent, brick slowly began to replace wood, and different building types (such as specialized stores, banks, telegraph offices and restaurants) began to appear.

In the East, the commercial growth experienced by cities due to transportation improvements and manufacturing developments, which eventually gave way to what later became known as the Industrial Revolution, was slowly beginning to alter habitats. Fireplaces, which up until 1840 had been used for cooking and heating, began to be replaced by cast-iron coal or wood burning kitchen stoves, not only for cooking but also for heating. Eventually gas stoves in large cities and kerosene units in smaller towns became standardized. Also, towards the latter part of the century, central heating furnaces and hot water or steam heating became popular and refrigeration was commonly achieved by the use of iceboxes.

Running water was not readily available in most cities, and central pumps and wells were the primary supply sources, although rainwater was also collected from the roof and piped into cisterns for storage.

Water was, of course, unheated; and although by 1840 boilers were being widely used, for most 19th century people bathing and washing were not daily routines. The weekly bath, although slowly becoming an established practice, was an inconvenient process since it usually took place near the warmth of the kitchen stove, with a privacy screen isolating the bather from the rest of the household members, who occupied whatever space remained available. The eventual development of a "bath room" by the end of the century, together with the general availability of hot and cold running water and sewer systems, alleviated most of these problems. Although the patent issued for the first water closet in America dates back to 1833, because of their dependency on municipal sewerage and water, it was not until 1900 that these units became available for urban dwellings.[4] By 1861 the functional and mass produced double-shelled enamel bathtub replaced the cumbersome cast-iron tubs with rolled rims and claw legs. And so the bathroom, with all its hygienic advantages, became a programmed standard in most American shelters.

Lighting developed from homemade candles into lamps using turpentine, whale oil, lard, train oil, stearine or other animal or vegetable products as fuel, as well as gas lighting (specially in the North). Eventually, refined coal oil (kerosene) came into use around 1860. In 1874, Edison patented the electric light, but it was not until the public production of electricity that its use became widespread.[5]

Slowly, as more and more industries were developed, machines began to replace manual labor and standardization began to set in. About 1850, machine-made furniture began to replace handcrafted pieces, and by about the same time mechanical washing machines, vacuum cleaners and dishwashers had already been conceived. Mechanization and technology had begun an unstoppable advance.

The industrial movement which pulled the worker away from the home and into the factory brought about a tremendous population influx from rural regions and from abroad into the sprawling industrial centers of established or emerging cities. The population of New York City reached almost two and three-quarter million inhabitants by 1890, 45 times more people than its population of 60,000 in 1800.[6] And although this rapid growth was mostly due to foreign immigrants, rural implosions were also a contributing factor.

Because of the overcrowding of urban housing units, the city's first tenements were built. Around 1833, the first purposely planned multifamily dwelling was developed, consisting of barrack-type structures and row housing, about 4 stories high, built in unoccupied spaces of poor neighborhoods and placed one behind the other or side by side along narrow alleys or in the backyards of old buildings.[7]

It was in trying to improve this type of housing that the "dumbbell" tenement was developed: a structure that ranged from 5 to 7 stories, with a 10-foot rear backyard and no front setback. Each floor was divided into 4 sets of apartments (14 rooms to a floor), with only the front sets of apartments receiving natural light or ventilation. Small air shafts (2 feet square), enclosed on all 4 sides and extending the full height of the building, provided the remainder of the units with a crude ventilation duct of sorts, though it was both inadequate and dangerous, considering that in those days the only means of artificial lighting was the use of oxygen-consuming open flames. Built until 1901, these tenements were to become the worst type of housing in the history of New York City.[8]

Public buildings in the eastern cities, however, followed a different pattern. In contrast to the lower standards of housing in many cities, most public and governmental buildings adopted grandiose proportions. It was during this period that capitols, libraries, colleges, courthouses and commercial buildings alike adopted neoclassical, Gothic and Romanesque styles. Towards the second half of the century, some significant changes took place with the introduction of skeleton cage structures and vertical transportation. In 1855 James Bogardus put up the first metal frame structure with brick infills, anticipating by some 30 years the general use of this type of construction, and in 1857, the 5-story Haughwout store building in New York became the first commercial building equipped with one of the most essential elements of the urban skyscraper: a passenger elevator.

The first public elevator building which took advantage of this feature for high rise office usage was the 7-story Equitable Life Insurance Company in New York (1868–1870), followed by Richard Morris Hunt's 260-foot-high Tribune building.[9] This newly discovered high rise concept was found to be extremely profitable with its small land base, adequate adaptability to choice locations and the availability of large net assignable spaces in the rapidly disappearing prime sectors of cities. And so the race upwards began.

MODERNISM

By 1900 most cities had begun to develop water and sewer systems, so fully equipped bathrooms became common. By 1918, cumbersome mechanical electric refrigerators were produced, but it was not until 1920 that more reliable gas or electric units became popular, replacing the traditional kitchen iceboxes. Also, by 1920 hot air, water or steam furnaces were considered standard in most dwellings.[10]

Communications also began to develop rapidly. The 48,000 telephones of 1880 were uncommon objects found only in businesses or in luxury residences, handling about 200,000 calls per day. In less than 100 years (1975) however, the number of telephones would rise to 149,000,000 units handling an average of 633,000,000 calls per day[11] making their widespread use a social necessity rather than a mere convenience.

As the 20th century unfolded, numerous technological and scientific advances contributed to further transform the spatial conception of American habitats. In the early 1900's, housing and public buildings were swept by new architectural and design conditions. The new design movements, which radically transformed every building type in the early 20th century, found their basis in the mechanization, technological development and scientific advances of the Industrial Revolution, the success of capitalism, and the moral ethics and religious righteousness of the times. Their general characteristics were formal simplicity, honesty and rationalism, disregard for the frivolous, or purely decorative, and an engineering-controlled approach to the design of shelters.

A striking contrast between earlier shelters and those of the 20th century design movements lies in the influence of regional and local characteristics on their respective spatial realizations. The standardization derived from the Industrial Revolution with its mechanical systems, environmental controls, and newly developed international styles of architectural design, eventually overpowered the influence of region, climate and culture alike, and unified, within a framework of technological simplicity, the fundamental value judgments employed in the development of habitats.

This also introduced new products and systems in building design and construction. Mass produced materials and design simplifications caused a decrease in individuality, lack of attention to detail and a relative loss of the human scale, as well as a dramatic redefinition of architectural aesthetics. Man set out to erect monuments to science and technology through his shelters.

Strangely enough, these revolutionary concepts did not involve drastic programmatic variations, but rather an overwhelming acceptance of traditional plan arrangements; they merely redefined and standardized the design of dwellings and building types on the basis of 100-year-old spatial layouts of Georgian origins.

Despite new and revolutionary design trends, the functional fragmentation of modern shelters followed growth and distribution patterns almost identical to those of the early colonists; and since then, the basic tendency has continued to be to enlarge or multiply building types as operations increase in complexity in the same manner as early colonists added lean-to's to the rear of their dwellings or separated kitchens from their halls and parlors.

In this respect, the functionalism displayed in the design of modern shelters has totally abandoned its utilitarian roots. The functional single-purpose area and the utilitarian multiple-use space are at opposite ends of the essential understanding of shelters. The meeting hall of the early colonies with its dual or triple functions was based on a concept that conflicts drastically with most contemporary shelters. One is very unlikely to find today building types that are suitable for conversion to functions other than those for which they were specifically intended. As it has developed from its humble origins, modern functionalism has unquestionably outgrown its utilitarian ancestry.

4
Groupings: Dynamics and Profiles

The political structures and social frameworks of the early colonists were those of their countries of origin. In those days, the settlers were not yet Americans, but a conglomerate of nationalities which gradually changed cultural heritages and idiosyncrasies to accommodate the characteristics and requirements of their new homeland.

The plans of villages or towns in the early years tried to duplicate the land division patterns of European countries, and it was not until later stages of growth that some began responding to local requirements and varying communal characteristics, and that the first indigenous patterns of development emerged. Thus to comprehend the origin of these roots, some basic elements of European planning must be identified.

Cities have traditionally exposed the basic socioeconomic and political characteristics of the groupings contained within them. All cities are conglomerates of streets and walls—streets which bring people together and walls that separate them, but the basic characteristics of an industrial metropolis or a retirement community, nevertheless, differ in conditions which far exceed the mere physical distribution of streets and walls. And thus settlements have been classified in a multitude of categories that cover from simple crossroad communities to large urban complexes.

It is very difficult, however, to find cities which satisfy any specific functional definition exclusively; thus what has traditionally distinguished a city from a primitive village is its higher degree of sociopolitical organization.

In the past, the two basic urban forms of permanent settlements were either walled towns or open cities. Historically, however, most cities did not start with a plan, but rather developed by accretion, and their growth was irregular and dynamic.

SETTLEMENTS IN HISTORY

Since early in history the growth of settlements has brought about the need to regulate urban expansion. The first city planning and building codes known to date were the codes of Hammurabi (1792–1750 B.C.), in which the punishment for irresponsible builders went so far as to have their offspring sacrificed should their buildings fall and injure the occupants.[1]

Except for a few exceptions (such as Alexandria and Rome) the size of the major early cities is easily dwarfed by contemporary urban conglomerates: Athens, in the 5th and 4th centuries B.C., had a population of 40,000 and Plato's ideal city size was between 5,000 and 10,000 citizens. When unchecked urban sprawl took place in the early years of Christianity, the lack of preparedness, planning guidelines and technical capabilities proved disastrous. Rome grew to a large and congested metropolis of well over a million inhabitants plagued with slums and crowded with 6- to 8-floor tenements. By the 4th century A.D. there were 35 times more blocks of apartments than single family dwellings in the city.[2] While city dwellers lived in slums, affluent Romans moved to country villas, following a pattern very similar to modern suburban sprawls.

Before their eventual transformation into larger cities, medieval towns reflected the social and political turmoil of their era through the use of encircling walls (thus protecting their settlements from hostile intrusions in a way very similar to that of modern private neighborhoods, apartments, or condominium complexes) but also maintained a communal tradition in which individuals never became a mass. In contrast with the rigid street gridiron of the Roman days, the medieval city had curved streets which helped cut down the wind and made the heating of houses easier. The city wall was the most expensive civic project and was designed both for security and beauty; this wall was the one feature that clearly distinguished the medieval city from a mere village.

Consistently, medieval towns retained the human scale and within their walls a high degree of communal participation was achieved. On the average their populations were made up of a few hundred people, with a maximum size of 50,000 citizens.[3] Because of their functional and utilitarian similarities, these were layouts and programmatic distributions which would greatly influence colonial communities in America later on.

Following these optimum medieval township sizes, increased world trade brought larger concentrations of people to those communities strategically located on main transportation crossroads, giving rise to what have become known as the neoclassic cities: during the 14th century, Florence doubled its population from 45,000 people to 90,000, Paris grew from 100,000 to 240,000 inhabitants and Venice reached 200,000.[4] Trade gained importance over the power of the feudal lords and a new class of wealthy merchants appeared. Growing populations, however, caused congestion in the cities. Seen in perspective, the problems faced then were not much different from those faced today by large metropolises. Behind luxurious facades, monumental avenues and grandiose plazas, dwelt a congested urban population. Cities lacked proper sanitary services such as sewers, water distribution and drainage; and epidemics, pestilence and poverty were slowly becoming major factors in an ever-widening gap between a privileged aristocracy and the impoverished urban masses. Migration to the newly discovered continent across the Atlantic increased; there, newly developed colonial towns and villages appeared to offer more room for expansion and better growth opportunities.

It was also during this time that a curious change began to take place: machines were slowly replacing handicraft methods in the production of goods.

AMERICAN SETTLEMENTS

At the beginning of the 17th century, almost all European powers competing for supremacy in North America were actively colonizing and planning new towns and villages in the New World.

Colonial expansion produced settlements in the New World that originally reflected the modest character, environmental needs and puritan simplicity of their founders, but later on copied the forms and patterns of medieval towns and villages and, in many cases, followed developments quite similar to those of their European predecessors. The settlements known as pueblos established by the Spaniards in America, for example, closely resembled the feudal patterns of land holdings in Spain. And French settlements in Canada and the Mississippi Valley were laid out according to certain elements (such as linear planning and gridiron layouts) common to their European ancestry.

Around 1610 new and varied English settlements began to appear in the New World. And as early as 1626 one already finds regulations covering town buildings.[5]

In the towns of the early 18th century, commons of approximately 300 acres surrounded a town proper with a central market place, lots for lords and proprietors, meeting houses and other public buildings such as courthouses, schools and storehouses. Sometimes a square house upon a hill served as church, but it also held six cannons upon its flat roof.[6]

Virtually all communities that emerged as the colonies grew were laid out on the basis of common patterns, since in New England the word "town" did not denote an urban style of settlement, but rather (in close similarity to the Spanish pueblos) an entire community of village lots and agricultural fields.

The sites for homes were generally selected near the center of the township, usually grouped around an open area where communal meetings took place. In general, most colonial villages developed around this central open space which served as market place, site of the meeting house and other public buildings and gathering place for cattle in case of Indian attack. Eventually, this single open square in the center of the town became the basis for the urban layouts of almost all American cities.

In these early settlements, building setbacks were carefully observed and the church generally occupied a place of honor facing the common. There was also a marked distinction between inner village and countryside. All, or nearly all, of the community members resided in the village but, in practice, the urban limits of many villages gradually increased as farm lots were divided into home lots.

These pioneer communities began as small fortresses and stockades enclosing dwellings which were little more than huts; architectural qualities did not begin to emerge until nearly half a century later. During this transition, however, building types and village layouts underwent drastic changes. Agricultural lanes were turned into residential streets and cultivated fields into house lots. The village green or common was reduced in size, sold or allocated for public uses, and building encroachments became more and more frequent. The siting of important buildings was also changed. Curiously enough, the simplicity of the village plans seemed to facilitate the adaptation to rapidly changing circumstances.

FROM VILLAGES TO CITIES. PATTERNS OF URBAN GROWTH

During the 19th century, most of the established American cities experienced high rates of growth due to increased migratory influxes, expansionism and industrialization. As trade and commerce thrived, cities grew; and as new streets developed, their sprawling grid patterns marked the beginning of modern urban checkerboards. Other geometries were explored; neoclassic influences appealed so much to the leading political figures of the time that they became determining factors in the layout of Washington, D.C. But in spite of a few such isolated cases, the gridiron pattern prevailed as the basis for the layout of most major American cities.

In 1811, after turning down a proposal from the city architect, an official commission composed of two lawyers and a surveyor proposed a rigid gridiron street layout for the island of Manhattan to be executed irrespective of topography or waterfront. In the first place, they assumed that traffic would move back and forth between the Hudson and East Rivers and thus created 60-foot-wide east-west streets 260 feet apart; secondly, by assuming secondary traffic flow perpendicular to the east-west direction, they outlined secondary streets in the opposite direction which, although wider (100 feet), were spaced at distances ranging from 600 feet to 900 feet. Only Broadway was retained as an angular avenue. Unfortunately, the commission's appraisal of the traffic flow was wrong, since traditionally New York's traffic has moved north-south. Furthermore, due to streets occupying more than 30 percent of the land area, economic justifications and land development compensations were made by sacrificing green areas. It was not until 1856 (40 years later) that the 840 acres allocated to Central Park were estab-

lished in the urban plan at a cost of $5,500,000 to the city's taxpayers[7]—
not a sum to be disregarded, considering that only 230 years before
Peter Minuit had purchased the entire island from the Indians for $24.00.

Many of the problems which seemed to have plagued American cities from the very beginning were not only social or technical in nature
but a result of political and economic factors as well. Municipal sewerage, for example, was firstattempted in Boston in 1652, but was inadequate from the start. It was not until 1857 in Brooklyn that the first
working municipal sewer system was established.[8]

Continuous and dependable water supplies were not available in
American cities until 1850; before that, individual pumps and wells
were the chief sources of water for most urban dwellings. Around 1900,
in a luxury hotel, one could get a room with a private bath at an extra
cost; but it was not until 1907, in Buffalo, that the first hotel was built
which provided every room with a private bath.[9]

The Industrial Revolution affected American cities dramatically.
Improvements in communications, transportation systems, and public
health and safety advanced more in 100 years than in all preceding
history, drawing people like fireflies into the glitter of exploding urban
centers; and suddenly economies began to shift from agricultural to industrial bases.

Throughout the 18th century the population of this nation was mostly
agrarian, with only about 5 percent of the people living in established
towns or other small communities. In 1790 only two cities had a population of more than 25,000 people. Yet these small-scale rural nuclei
were not to survive. By mid-century, 20 percent of the American people lived in urban communities. From that moment on, the increase
accelerated: in 1940 there were over 3,400 urban communities in this
nation harboring 56.5 percent of its total population; 412 cities had
crossed the 25,000 inhabitants mark, with 23 of these having between
250,000 and 500,000 people, 9 between 500,000 and 1,000,000, and
5 having already exceeded 1,000,000 urban dwellers.[10]

With the railroad and other means of transportation also facilitating
travel to and from factory towns (which eventually developed into industrial metropolises), and the large migration to sprawling work centers, the increased number of tenements and other inadequate housing
arrangements accelerated urban decay.

This should not be interpreted as the beginning of all urban problems. The Industrial Revolution merely served to hasten an otherwise

inevitable trend of most large settlements. Slums were not a direct product of industrial growth, but rather of a natural pattern of urban overcrowding which has always plagued societies. Babylon in all its 6th and 5th century B.C. grandeur was lined with the 3- and 4-story tenement-type dwellings of the populace.

Obviously, increased migratory influxes could not be assimilated indefinitely without expansion. As cities grew taller, so did they sprawl and ramify horizontally. At the dawn of this century, the high costs of land squeezed the single family dwelling out of the inner city and into the outskirts. These suburban sprawls reached such distances that facilities which once had served entire urban populations were no longer readily accessible by the distant areas, thus creating the need for the development of new facilities and services in suburban locations and giving them the character of satellite communities. And so the circle was closed. Decentralization had begun.

Although suburban expansions were prompted by exploding urban populations, the excess population from the central cities was not entirely drained, so with work centers changing continuously and people wanting to live near their places of employment, the tendency to shift from one job to another evolved into moves from city to city, and thus the urban population became transient. This was a new characteristic for which freedom of movement was a necessity that could be satisfied only by such accommodations as the rental apartment. In addition to the 19th century tenements, new multifamily dwellings for higher income families began to sprout in every American city. The urban nucleus became a large work center where families proceeded to hire apartments for temporary occupancy rather than dwell in a permanent home they owned themselves.

Since World War II the population explosion and improvements in communications, transportation and living standards have greatly increased suburban developments, resulting in the proliferation of small satellite communities with inadequate resources to administer many of the "urban" functions (law enforcement, fire safety and public works projects) demanded from them, since many of these municipalities fall outside the jurisdiction of larger (or wealthier) urban sectors or counties. This is especially notable around large urban centers: the metropolitan Chicago area has well over 1,000 local communities, 600 special-purpose governmental units (school, park and sanitary districts, etc.) and some 400 different municipalities.[11]

The most common pattern of growth of American urban nuclei is as follows: accelerated rate of development; lower densities in new residential, commercial and industrial sectors; and a tendency towards urban megalopolises, where adjacent urban and suburban sectors blend into one large "metropolitan" nucleus, which sometimes adjoins and extends into other "metropolitan" sectors, thus producing a full string of megalopolitan units (Boston to Washington, D.C., Chicago to Milwaukee, Los Angeles to San Diego, etc.).

However, because most of the job opportunities remain in urbanized sectors, as cities spread into rural areas, the population profile of the American city during working hours resembles a pyramid from which a complex web of communication arteries extends outward into the countryside, and is thus strained between the centralizing and decentralizing tendencies that have rarely had the opportunity of developing within the framework of proper planning.

Despite an increase in self-contained community buildings and other inner-city renewal movements, most urban dwellers have continued to consider the city a work center, not unlike workers during the Industrial Revolution. In this sense, there has been no real social change whatsoever. It has been accurately observed that, as presently constituted, most American cities are extremely wasteful, emptying themselves day after day into suburban developments at night and further into countrysides every weekend, a trend which has made large urban communities and the shelters they contain highly impersonal and significantly unstable.[12]

The pattern of development followed by cities is inherently linked to the social order of their inhabitants. The pharaohs of Egypt and the emperors of Rome erected huge projects dedicated to the glorification of demigods or mighty rulers. The Greek cities, on the other hand, were planned for people. Human scale was the primary measure in the urban conception of the Hellenic city. Likewise, the medieval town dwelt within the parameters of its encircling walls. Social and religious gatherings, exchanges and commerce occurred daily in its plaza and provided common bonds for its dwellers.

The American city was born of this latter tradition. Yet, because of its exploding development, many of its original characteristics changed and urban conglomerates were filled with ever-larger projects built in tribute to technological developments and industrial growth.

The impact of the Industrial Revolution was felt in cities throughout the world in different ways, but inevitably with similar results. Geometries and patterns in city planning were originally introduced in accordance with the structure of the land or imposed by ruling authorities, thus clearly illustrating a nation's sociopolitical structure. For people raised on American gridirons, for example, a place like Japan is bewildering. Americans name the communication arteries, while the Japanese name the intersections. Thus, in a system which emphasizes hierarchies based on time, Japanese homes are numbered according to their age, and not according to their relative position along a specific street. In a Japanese neighborhood, the first house built will be house number one and a constant reminder to the residents or visitors to the area that it was placed there prior to house number twelve.[13]

Likewise, many urban patterns also facilitate the development of cultural behaviors. The constant social contact of Latins in their urban plazas contrasts with the lined-up main street activities so characteristic of this nation.

Yet industrialization has forced many of these cultural differences to give in to new spatial adjustments. Precisely because of its emphasis on intersections rather than communication arteries, Japan has had extensive problems trying to integrate the automobile into its urban patterns, and has suffered some of the world's biggest traffic jams in its large cities. And in Caracas, Venezuela—a country which is only beginning to industrialize—rush hour traffic is so thick that the government has banned each private car from the streets one day a week (depending on the car's license plate number), trying to achieve an approximate 20 percent reduction of its traffic congestion.

Of all the legacies of the Industrial Revolution, the automobile, in particular, happens to be one of the greatest consumers of urban space. In 1895 there were four automobiles registered in the United States; by 1900 the figure had risen to 8,000. Less than 80 years later, in 1972, there were 97,000,000 registered cars in this nation.[14] In Los Angeles, the space devoted to cars and their related spaces (streets, freeways and parking) amounts to approximately 70 percent of the city area.[15]

The consequences of this type of urban development tear the city's fabric apart. Physically, people lose their incentive to walk or simply cannot find *where* to walk, while also being plagued by a multitude of other health-related hazards. And psychologically, motorized vehicles affect human motion through space and its temporal correspondence. Man can move comfortably at a walking speed of 3 to 4 miles per hour. To appreciate older buildings with all their elements in proper scale, one must walk around and through them; actually, *that is the way such buildings were designed to be seen.* Today, the only justification for the large even-textured facades of contemporary architecture is the radical kinesthetic change to which human beings have been subjected. Modern building complexes with their overwhelming collection of pre-manufactured architectural elements are aimed at a much faster visual perception than their classic predecessors. Looked upon closely, they become endless exhibits of repetitive planes and materials.

It is noteworthy, however, that despite rapid changes in urban structures, the replacement or upgrading of buildings and utilities within urban nuclei occurs only gradually. Cities consist of a combination of past and present conditions and consistently exhibit a mixture of new technologies applied to modify past conditions in response to changing cultural and economic factors.

5
The Process of Modernization

Since colonial times the essential characteristics of American habitats have been steered by modernization. Because modernization is a direct function of social transformations and technological progress, it can be defined as the cohesive development of socially interactive institutions in conjunction with unfolding technoscientific advances.

The early colonists were able to develop their utilitarian habitats with little external aid; but later, due to increased occupational fragmentations and the diversification of architectural styles, most households became incapable of developing adequate shelters without specialized help.

A strong interaction between economic development and technological advances has defined the evolution of American shelters from the outset. America never experienced Europe's great transition to industrialization: feudalism. The brief transition between the 18th and the early 19th century was the only period in which the United States could be said to have undergone a development reminiscent of that of the Middle Ages. However, this period was so brief that the gradual transition from the "medieval" city to the modern industrial conglomerate was never truly experienced. In 1790, most Americans lived in rural areas (the entire national population was only slightly larger than that of Portugal) which were poorly connected: a ton of goods could be brought from Europe for $9.00 but it could not be transported on American soil for more than 30 miles for that same amount of money.[1] Yet within the short span of 50 years there were revolutionary improvements in transportation, including the development of inland canals, steamboats and stagecoaches; and eventually, the most explosive growth in railroad communication the world had ever known.

Transportation was one of the essential pillars of the economic growth of the 19th century; from the very beginning all aspects of the North American socioeconomic development depended on the adequate distribution of its plentiful resources. The journey up the Mississippi River in the days prior to the steamboat was so difficult that a trip from New Orleans to Pittsburgh could not be made without a crew of strong men and, even then, it took at least four months to complete. In 1807, however, steam-powered boats began successfully sailing up the Hudson River. Railroads followed: the national mileage of railroad lines grew from 72 miles in 1830 to 30,603 in 1860.[2] Technology had begun to close gaps in the North American quest for modernization.

ECONOMIC AND TECHNOLOGICAL INTERFACES

The American economy before the middle of the 19th century was still not strong or secure enough to liberate itself from the ways and customs inherited from Europe. These first decades of colonization are characterized by a gradual economic growth which started with the exploitation of readily available resources using inherited systems, which were sometimes altered to include indigenous technologies. Soil characteristics and weather patterns were the only determinants which defined staple crops.

In a similar fashion, the main driving forces behind the development of habitats were mostly guidelines established by heritage, only slightly modified to comply with territorial and climatic conditions.

Land development followed this course at first, but not for long. Plans of collective, company-owned organizations failed due to the lack of productive work by laborers who knew they would be taken care of one way or another and thus originated one of the most significant factors in further American territorial developments: private ownership of the land. As early as 1618 the European system of land ownership by a company for which laborers worked, keeping only part of the crop production, was modified in Virginia to allow for private land owner-ship. From very early in history, land became the basis for social sta-tus in the New World.[3]

Hence, private ownership of land became a typical trait of the New World, and land rental (when adopted) was merely regarded as a tran-sition to ultimate proprietorship. As a result, dealings in land became widespread and many farmers and merchants became real estate spec-ulators. Land dealings were, in fact, rampant during this time, leading many settlers to abandon their shelters, farms or communities after cy-cle crops were harvested or land speculation deals were completed. During the expansionism period, colonial settlers were not as "stable" as many choose to believe. Actually, throughout this period housing—especially rural housing—was not the chief speculative real estate commodity: *land* was.

Through 1800 the major sectors of economic expansion were in the areas of foreign trade—shipbuilding; and in domestic commerce, transportation and manufacturing. With a more diversified economy, regional differences began to surface: the leading population expansion of the early 19th century took place in the great port cities—Balti-more, Boston, New Orleans and Philadelphia—which were main cen-ters of access and egress to and from the American continent and its trade markets. Not surprisingly, it is in port cities that the first signif-icant influences of European architectural styles and explosive urban developments and problems are found. Despite the fact that Boston was the first American city to install a sewer system, as late as 1930 half of the city's population lived in either 4- or 5-story brick tenements or in wooden "three deckers" (3-story wood frame structures which housed three or more families).[4]

As a result of railroad expansionism and improvements in commu-nications, land speculation and real estate ventures intensified. Throughout the West, promoters or investors would buy inexpensive land acreage from the government ($1.25 an acre) after analyzing maps of projected railroad, post roads or water routes and establishing likely spots for the development of new towns. All that was necessary to make the investment succeed was to persuade the railroad to build a station in those locations.[5]

Residential development followed a similar pattern, with investors acquiring inexpensive land for labor housing, mortgaging the land to finance a plant which would employ those laborers, and eventually increasing the land value once the housing was occupied, thus establishing an investment which profited from manufacturing *and* real estate improvement.[6]

It was not only the pursuit of profits, however, that lay behind increases in land values, but the patterns of development which emerged as a result of the exploitation of America's vast resources and the growth in inland commerce. The 14 counties bordering the shores of the Erie Canal showed an increase of 91 percent in land value while the rest of the urban sectors in those areas showed only a 52 percent growth. Similarly, employment in manufacturing and commerce grew more rapidly in those same 14 counties than elsewhere in the country.[7] Cities like Buffalo, Cleveland, New Orleans and Chicago were unquestionably stimulated by their connections with inland water transportation channels. And thus, with increased land values and urban implosions, public or commercial buildings and land developments began to show the first typical symptoms of unchecked urban growth: multifamily units, smaller lot sizes, mushrooming building construction, sanitation and public works inadequacies, etc.

As commerce increased, new communal structures and urban patterns began sweeping the nation. Ultimately, the pattern of development followed by many towns built around railroad stations did not respond to the local conditions, but rather to the preconceived formats of the investor or the newcomers.

That is also why today most railroad stations are located at the epicenter of urban conglomerates, since the original crossroads and ports of call that eventually developed into larger cities evolved around these stations. The larger the city grew, the more it needed to extend its communication arteries, and today one can easily equate the size of an urban conglomerate with the complexity of the railroad webs that cross it. Paradoxically, modern urban traffic is being hampered by its own antecedents.

Curiously, although America was only partially explored during the 1800's, its undiscovered landscape had begun to be widely altered by communication scars before the turn of the century. And thus, without reluctance or reservations, America entered its man-made era.

This new wave of machinism led to the enormous proliferation of a building type that until then was practically unknown: the factory; and eventually to the industrial city. The drastic impact of this development altered housing, building types, transportation, life-styles, schedules, public services and almost every area of urban endeavor.

The 20th century marked the beginning of an organizational revolution in the North American economy. The consolidation of business and the development of capital markets facilitated the rise of corporate capitalism. Iron and steel ranked number one among the industries in 1900 and second through 1920. The automotive industry, however, was not to be denied; its dramatic rise measured in value of products rose from 150th place in 1899 to first place in 1925.[8]

This rapid expansion of the automotive industry caused drastic changes in American urban landscapes; as early as 1923, the first four-lane parkway with landscaped median was completed in New York. It was also this explosive automotive expansion that caused a significant addition to the typical American housing unit: the garage. This space was originally developed as an addition to existing shelters, and it is noteworthy that garage construction—as a result of the proliferation of the private passenger automobile—started only about 60 years ago. Thus, the garages of single family or multifamily units built around the early part of the century are detached structures (almost invariably constructed after the original shelters), generally placed at the rear of the property (obviously equated to the carriage and horse barn).

One significant growth sector of the American economy throughout the early part of the 20th century was electric service, and with it, the proliferation of new electrical machinery, tools and appliances, all invading every corner of American built environments. It was a period in which the nation was enamored with technological innovations, from airborne bicycles to mechanical "house shinglers."

It was also during this period that rapid mass transit systems began accelerating the economic development of urban conglomerates. The city became a formidable consumer as well as a formidable provider. As early as 1887 Andrew Carnegie had already shifted the production of his foundries from rails to structural steel, and, upon the arrival of the 20th century, American cities already constituted more important markets for the steel industry than the railroad. The mushrooming urban conglomerates were absorbing massive quantities of steel and every other available construction material.

If there ever was a birth of the American construction industry as such, it must be placed within these early 20th century years of urban

implosions, transportation improvements, and explosive technoeco-
nomic developments. By mid-century, continental distances were reduced tenfold and cities had become machine-made complexes built
by socioeconomic empires.

Although the economic growth of this nation was seriously shaken
by periods of economic recession and depression, during and after World
War II America was to see one of the sharpest rises in building construction in history (wartime housing and factory expansionism being
among the most significant).

During the period 1946–1950, immigration increased approximately
73 percent from the previous decade and permanent dwelling units
doubled.[9] This renewed immigration influx, which by the early 1950's
represented a 20 percent increase in this nation's labor force, added to
the factors which caused the postwar housing boom.

New technologies also contributed to the improvement of building
construction, such as the development of corrugated aluminum and
aluminum windows, heat-absorbing and glare-reducing glass, tem-
pered glass and the architectural curtain wall, as well as the use of
plastics and synthetic materials. The use of prefabricated parts also became widespread, and this was especially important in the construction
of housing units, because of the savings realized by the standardization
of furnishings and equipment.

The American economic growth in the 60's was spurred basically
by housing construction, with an approximate 34 percent growth experienced from the mid–50's to 1970 in the total fixed capital devoted
to it, but by the end of the decade the explosive energy consumption
coincided with a decline in American gas and oil reserves, causing an
increase in oil imports which eventually led to the exorbitant rise in oil
prices of the 1973–1974 embargo.

Because of this, the price revolution of the 70's caused a significant
shift in investments, income distribution, rate of inflation and balances
of world trade, and set in motion the severe recession which took place
in the middle of the decade. In this country few companies, businesses, or industries related to housing, the building trades, or construction in general escaped its extreme hardships. Many failed. The
questions remain, with a slow recovery through the first half of the
80's: when and how will the American construction industry regain
economic prosperity? Or even more fundamentally: will it ever?

A HERITAGE OF PARADIGMS

Since the early days of the colonies, the development of shelters and settlements in this country has been characterized by distinct patterns of growth, fragmentation, and territoriality which have consistently interacted with the resource availability, economic conditions, technological advances and cultural heritage factors of specific times.

The patterns of growth in shelters were originally structured by utilitarian characteristics, but soon gave way to socioeconomic influences, and the simple room additions of parlors or kitchens eventually became complex spatial arrangements where the household functions began to be isolated and localized.

The patterns of fragmentation followed by shelters are closely related to the increasing functional complexities of evolving societies. The multi-purpose log cabin of the colonies and the sauna room of a modern residential complex are at opposite ends of the spectrum of utilitarian/functional spaces.

The early parlor/hall fragmentation marked the beginnings of a trend towards shelter permanence in the New World. The conflict between increasing social contacts and the need for household privacy forced this primary fragmentation. It was a process which was justifiable only when the acceptance of community life, contacts and ties became sufficiently engrained in the people's attitudes and behaviors. Yet, as further spatial allocations took place, the fragmentation patterns came to be based more on cultural behaviorism, increasingly influenced by social models and symbolic of economic class.

Society (and American society above all) has consistently linked the spatial fragmentation of its habitats to economic affluence. In modern "luxury" shelters, even the basic programmed spaces experience further fragmentation: a bathroom is partitioned so as to isolate lavatories, dressing areas, water closets and bathtub areas (sometimes even adding a separate shower stall); kitchens grow "breakfast nooks" on their sides, and even living rooms contain sunken "conversation pits." Opulence and functional specialization have gone hand in hand from the beginning. It is not surprising, therefore, that as economic environments prospered, so did the complexity of their spatial layouts.

As modernization developed, these fragmentation patterns were extended to stylistic considerations, thus establishing the basis for building prototypes. Even today, there is a strong linkage between classicism and public institutions. The arch, the buttress, the Corinthian order and the Georgian house have all acquired their own symbolic significance and are presently identified accordingly.

The interaction of territoriality and cultural heritage was also an obvious influence on American habitats from the outset. Consistently, the early colonists tried to re-create their cultural surroundings and only deviated from this goal because of territorial demands.

In the late 19th century the social, economic and technological impacts of the Industrial Revolution altered all of these interactive patterns drastically. Movements or events that affect the interaction of these patterns of development result in significant transition periods or historical "revolutions." This was the case with the industrial growth of the late 19th and early 20th centuries.

The original concept of shelter was reshaped to allow for temporary occupancy, thus modifying one of the most basic characteristics of a "dwelling." Until then permanent housing had been conceived with long term plans in mind and a strong sense of permanence. The rapid industrialization and urban growth of the early 20th century reversed these patterns, unsettling households and shelters alike. Eventually, houses and buildings were serialized; and by the middle of the century, large development corporations were mass-producing entire neighborhoods. This serialization concept, born out of the industrial assembly plant and elevated to the category of fundamental building process, eventually engulfed the human shelter.

Added to this were the natural complexities of new industries and developing communication arteries, exploding scientific advances, and the divergent economies of the different national regions. This dramatic growth of building types from colonial days until modern times is illustrated in Figure 1.

Patterns similar to those followed by the development of shelters are perceived in the development of settlements, with easily distinguishable growth, fragmentation and territorial characteristics.

The patterns of growth followed by settlements have depended greatly on socioeconomic factors, transportation, communication, natural resources, etc. Yet the gradual sprawl of urban centers throughout the 18th century and the urban expansions of the second half of the 19th century cannot be considered equal models of development. The first stage of urban growth was characterized by gradual increases in housing, services and building type fragmentation together with a moderate

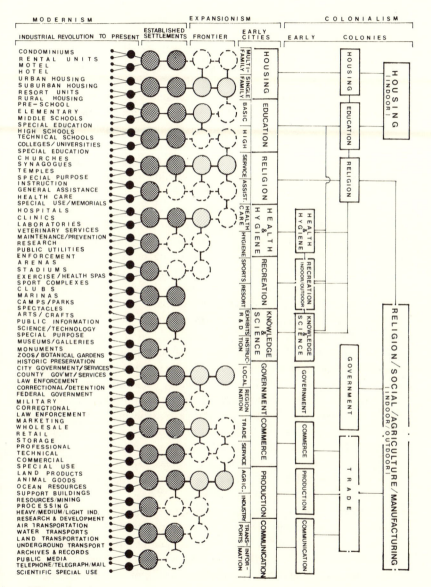

Figure 1
The Fragmentation of Building Types

population growth. And actually, in some cases, such as in the mining regions, the patterns of development even followed a negative course, since the fate of these settlements was strictly dependent on the availability of a specific natural resource for exploitation. The urban implosions of the Industrial Revolution, on the other hand, were characterized by the uncontrolled sprawl of tenements, industrial towns and communities; by the origination of satellite towns; and by a drastic modification in the essential basis of the urban conglomerate: the transformation of traditional cities into huge work centers upon which rural populations converged. They also brought about a dramatic change in the stability of the urban dweller, converting him from a permanent to a transient or temporary inhabitant.

The distribution of this urban growth also underwent dramatic changes. The exploding populations of the 19th and early 20th centuries affected small communities and large settlements alike. Industries sprawled uncontrollably from urban to rural areas, regulated only by communication channels or the availability of labor. Because it was densely populated and had better communication networks, the northeastern United States became the industrial region par excellence.

Varied urban distributions and occupational specializations combined with increased urban implosions resulted in strong socioeconomic fragmentation patterns, in which inner cities, ethnic sectors, suburbs, commercial and industrial areas, slums and high rent districts began to become isolated from one another. These divisions generally created cycles of boom and decay for almost every neighborhood, each cycle taking anywhere from 70 to 100 years. Housing sectors, built at ever-increasing distances from inner-city cores, went from high income to low income, and so did their corresponding commercial districts, service buildings and recreation areas. Today, this decay already has reached many former middle income or even high rent suburban neighborhoods in most cities.

Territoriality and location have also been essential factors in the development of urban nuclei. From the stagnating town bypassed by the railroad or major highway, to the distinct and peculiar characteristics of the harbor city, location has traditionally affected the profiles of cities. During the Industrial Revolution, this factor became especially significant due to the sprawling communication networks. Nowadays, further improved means of transportation and communication may be contributing to a reduction of the full impact of territoriality on urban nuclei. What the Industrial Revolution did on a regional basis a century ago, the transportation and communication advances of the present electronic era may be doing on a national basis today.

Presently, a combination of technoscientific advances, environmental concerns and changes in availability of resources may already be causing another significant transition period similar in nature to that of the Industrial Revolution. The availability of resources will affect the growth patterns of dwellings and settlements drastically; houses and buildings may get smaller, suburban areas may stagnate and so on. Computer usage and environmental concerns will also affect the present patterns of territoriality and fragmentation. Immediate communication is no longer limited to distance. Urban sprawl is being challenged, and many new scientific discoveries as well as several traditional values are being seriously questioned.

Contextually, America's quest for economic growth and technological advances has produced the greatest rise to affluence and social egalitarianism experienced by any nation on Earth. Today, the difference in income levels among social classes is the smallest in history. In fact, from 1972 to 1977 the population living below the official poverty line level (corrected for inflation) actually declined 0.3 percent. Federal personal income taxes decreased proportionally 0.4 percent from 1972 to 1978, while in that same period the real per capita disposable income rose 17 percent.[10] At the start of this decade, and after taking into account population growth factors, taxes and inflation, the real incomes of Americans have gone up, not down.

Housing and buildings in general reflect this socioeconomic reality in a multitude of ways. The middle-class home or the small commercial building of today has more amenities than the house of the rich or the large office building of the 1920's. Within half a century the spatial amenities and quality of American housing have bridged gaps thought impossible to conquer less than a century ago. Hence the devotion to science and machinism. The American rise to affluence has resulted in a world in which everything desirable is technologically possible and everything technologically possible is desirable.

Today many question technoscientific achievements and argue that their consequences cannot be easily managed. For example, attempts to control the explosive development of the automobile have met with little success. Other observers point out that the power of scientific progress is being overestimated by a machine-dominated society which believes that the world is more predictable than it really is. Undoubtedly, the maladies of urban nuclei have been greatly influenced by uncontrolled technological developments; but in fact, the success of the American economic system is a result of these same developments.

Progress is a self-sustaining force. All scientific achievements induce and produce new requirements and developments themselves. The artificial modification of ambient temperature used to be considered remarkable, but now there are levels of "comfort" which must be achieved. As more perfection is attained, more perfection is demanded. To cool the environment evenly 10° below the outside temperature is no longer enough; comfort levels of specific degree readings and percentages of relative humidity must be attained *no matter what the temperature is outside*. The shift has been from tampering with temperature differentials to the actual attainment of "artificial climate." And thus modern man's physical and psychological tolerances of climatic changes have been significantly reduced.

The blessings of modernization demand payment from society in the form of increasingly complex cultural and physical adjustments. Modernization is inherently linked to social transformation and technological advances, but its rate of progress is accelerated or obstructed by the human condition and its collective implications. In the final balance these are the true powers that be.

6
The Human Condition

The building of shelters is not a uniquely human activity. A hermit erecting a building to be inhabited only by himself would be a fabricator just like members of many other animal species. What is characteristic of man is his condition of social being. Human built environments reflect this fundamental facet in a multitude of ways.

The human condition is essentially made up of both collective and individual traits which are the foundation for all social structures. Historically, the growth of communal settlements and public affairs always occurred at the expense of families and individualistic expression; and the restrictions on the invasion of the private lives of citizens during the emergence of the first city-states was not based on a respect for private property as we know it, but on the need to allow each household unit a permanent location from which to participate in the public affairs of the community. In Greek cities, for example, citizens were obliged by law to share their harvest with one another, but each one was the sole owner of his land.[1]

One of the basic characteristics of the household unit is that of being a group of persons living together because they have either family ties or common patterns of wants, individual interests and general needs. Social groupings, on the other hand, are characterized by a greater freedom of choice and association, and have traditionally been based on the equality or similarity of their members. And these basic characteristics constitute one of the first significant differences between communities and households, since the latter have traditionally been social units of an accentuated inequality.[2]

The interactions of these basic components of the human condition—individualism and collectivism—are among the most powerful determinants of man's built environments. For an individual, single shelters offer a better opportunity for personal expression than the openness of public places. Colors, patterns and textures widely used on interior surfaces in houses and buildings are unthinkable finishes for external usage.

In essence, therefore, the human shelter becomes a mixture of the personal input of its dwellers and the social influences and restraints acting upon them at a given time.

If one analyzes the development of these individual and collective influences presently at play, however, some basic conflicts between subjective traits and collective influences become obvious. On the one hand, there is a wide selection of styles; but on the other hand, social egalitarianism leads to the endless repetition of "models" and "prototypes," that restrict or standardize building areas, functional distributions and construction technologies throughout the nation.

On a communal scale, socioeconomic levels cluster clearly identifiable groupings by locking "Joneses" within determined urban sectors. Because of this, the urban and suburban areas of most cities maintain a constant struggle for higher or lower density ratios; and consistently, the issues of low rise vs. high rise projects, traffic patterns and buffer zones become critical in new construction projects. Here, the "individual" expression becomes that of a community within an urban conglomerate, a community which, in effect, develops characteristics very similar to those of individual dwellings within neighborhoods. There is an inherent correspondence between well-kept gardens around individual shelters and the overall landscape appearance of entire neighborhoods, between the size of lots and the width of streets, etc.

That being the case, changes at the collective or the individual level are judged within the same contextual framework in matters concerning real estate values: residents belonging to a lower social stratum are as undesirable on a street as low income housing projects are in a middle-class sector. Even townhouse developments (although it has been proven that they *do not* damage the economic status of most neighborhoods) are generally considered one of the first symptoms of decay for an area since they increase urban density and are never evaluated for what they really are: a direct economic result of the natural overpricing of the community's land.

An important aspect of the human condition and its relationship to the evolution of habitats is how shelters themselves affect the people who inhabit them and how urban or rural settings affect their dwellers collectively.

Urban environments are known to cause considerable psychological stress in their dwellers, contain high levels of air and noise pollution, have the most alarming ratios of crime and social conflict, and involve a high cost of living. These disadvantages, however, are not critical enough to oust the typical urbanite, who finds significant appeal in the higher quality and more varied cultural and entertainment resources offered by the city, better job opportunities, increased privacy and individual anonymity, and better public and private services.

The differences between communal settings eventually translate into differences between collective groupings. Urbanites, for example, have been found to be less gregarious and outgoing than people in rural settings. And communal involvement and the individual sense of permanence also decrease with urban growth.

Further effects of the spatial setting on the human condition are also reflected in the behavioral differences within defined urban sectors. In a student housing complex at the Massachusetts Institute of Technology, researchers found that in buildings located around U-shaped courts, the people living at the open ends of the court, facing the street, had half as many acquaintances as those living within the courts themselves obviously because of greater sense of community in the latter setting. And after extensive research was done in a St. Louis housing project, it was concluded that the main reason for social maladies ranging from litter and vandalism to criminal offenses was the lack of semi-public areas that would facilitate the gathering of neighbors and thus encourage human interaction.[3]

The subjective interpretation of space is also a highly significant characteristic of the human condition which has traditionally affected man's relationship to his shelters. The child, for example, perceives the world on a much larger scale than his/her elders. The relative dimensions of objects are nothing but the subjective appreciation of their size in relation to oneself. One may measure the width, length and height of a room in feet, inches, meters or centimeters, intellectualize the meaning of measurements and communicate them to another in this fashion; but the *true* conceptual measure is done in steps, arm lengths and intimate relationships. The child might remember a very high ceiling in a room because it could never be reached, even by jumping on the bed; but in fact the room probably had a regular 8-foot high ceiling. It is natural for the subjective adjustment to wide rooms, high ceilings, large corridors, etc., to remain with children until adulthood. Most people who have the opportunity to return to their childhood home experience a certain degree of scale maladjustment and proportional reevaluation. "I remember it being so much larger," most will say. Those trying to reproduce their childhood home from memory will inevitably overestimate its size. In effect, the child's spatial world is a dynamic complex of relative limits that does not stop shrinking until adulthood.

Moreover, nowadays the majority of shelters are not developed by their inhabitants, as they had been traditionally, but by a speculative process in which programmatic standardization forces most households to interact within similar spatial confinements. A very minor percentage of the population has actually had any input whatsoever in the spatial distribution and proportions of its habitats or in their shelters' internal or external appearance.

Because of this, modern man must continuously adapt himself to new spatial distributions and sizes. The functional adaptability to preexisting spatial arrangements is a relatively recent social demand—imposed upon man within the last century or so—that has become essential to modern society. And since spaces cannot be readily redistributed, modern dwellers end up with the curious paradox of having to rearrange individual functions to accommodate themselves in rooms which were designed with a collective—rather than individual—concept of *functionalism*.

In effect, social mobility and the concept of spatial functionalism are in conflict with one another. And this conflict, in turn, increases the lack of permanence and attachment to the dwelling. Shelters become a mere commodity to be lived in, and are perceived as an "investment" rather than a "belonging," whose maintenance and remodeling expenses become a function of economic feasibility, marketability and life cycle costs rather than individual desires, household needs or just personal pride. Because of this, modern man is not only influenced by impersonal buildings, but in many cases ends up creating impersonal buildings himself.

These patterns of collective standardization, however, were not always present but have slowly evolved from a time in which the relationship of individuals and families with their shelters was quite different from today. It is on this transition that we now focus our attention.

THE FAMILY IN TRANSITION

One significant characteristic of the communal groupings in the early colonies was their family-centered social structure. The most meaningful cultural factors found in the colonial families were their specific origin, their attitude of acceptance or rejection towards the values and convictions of their homelands, and the changes that the frontier conditions and demands wrought upon their essential structure. For example, although the families who settled in New England were originally patriarchal in character and tradition, the status of husband and wife changed soon after they arrived in the New World due to the need for a newly expanded role to be filled by the woman—gardening for edibles and manufacturing of all clothing besides maintaining the household in general—since the man found his time occupied with building the home, clearing and plowing the land, hunting and, in general, providing for the family.

During the latter part of the 18th century however, the beginnings of the Industrial Revolution began to change the essential role of family members and the social interaction of families with the community. And except for some rural areas where traditional household units remained unchanged, most of the industrial functions of American society were transferred from the home into the factory.

Since agriculture was the primary economic support of this country, and was still a highly profitable occupation, the men remained involved mainly with agricultural activities; thus much of the factory work went to women first, and later on to the older children. In 1831, 68 percent of all employees in the cotton industry throughout the country were women, and in Massachusetts 80 percent of the cotton mill workers were females.[4]

The vacuum cleaner, the electric washer and dryer and other newly invented household appliances were not commodities which sprang from the mere desire to ease women's household chores, but rather from the necessity to liberate women's time for industry and factory work.

What this change in the role played by the wife and children did in the 19th century was to modify sharply the basic structure of families and communities throughout the nation. New relationships developed among family members, the authority level of the husband was greatly reduced and the essentially patriarchal structure of the colonial family no longer found sociological grounds in which to flourish.[5] Gradually, society became less family centered.

The roots of this nation's present family structures lie precisely in this specific period of time. Most of the significant changes which have taken place since then, such as the development of mass education, urban implosions, equal rights for women, increased work force mobility, mass communication systems and the enormous expansion of government interference in private affairs, cannot be credited with the same degree of influence that those basic movements of the 19th century had on the family unit.

However, in this light, there is one observation which must be added: none of the changes which have taken place since the Industrial Revolution have contributed in any way to *narrowing* the family gaps created by it; on the contrary, they have contributed to *widening* them. Curiously enough, it was about the end of World War II that a larger number of new families, technological advances, migrations to the sunbelt regions and the subsequent need for new housing developments (some of which did not include basements as traditionally developed in the North) were the primary factors behind one new—and characteristically American—spatial allocation in housing units: the *family* room. Varied family roles, and the need for a flexible space which would accommodate games, hobbies, and newly available household entertainment equipment such as radios, phonographs and television sets, were the main reasons for the inclusion of this specific room in the basic program of American housing units. Over the years, the family room has evolved to become alternatively a game room, study, den or guest room, library, socializing and eating area, or even a hobby shop. It has also slowly relegated the living room to be used for progressively fewer functions.

"Family room," however, is a misleading term, because its functional conception never implied its use by the family members *concurrently* only, but alternatively or sporadically as the case may have required. It was never intended to be a room which would join the family members in family-type activities exclusively, but an area to accommodate new amenities and equipment available for household usage for which there had not previously been any specific location within the confines of the average shelter. In this light, the family room is strictly a consequence of the technological revolution of the 19th century and never contributed more or less to bringing families together or keeping them apart than did the radios, phonographs, television sets and games contained within it.

DYNAMICS

One important issue which must be defined prior to analyzing the adequacy of shelters for modern family structures is the basic difference between *family* and *household*.

A distinction between families and households was recognized even in the first American colonies. Sometimes, for instance, due to the lack of proper housing, single males were assigned to live with certain families, thus modifying the household membership to include someone other than a family member; and in many cases—either through the addition of servants, in-laws, or other relatives—several individuals remained within the household conclave. However, if during colonial times a general distinction could be drawn between traditional families and other household units, this would soon change, since during the Industrial Revolution the variety of family structures and household types proliferated enormously.

Presently, the United States Census defines a *household* as a group of persons occupying a house, room, apartment or other group of rooms regarded as a housing unit; and a *family* as consisting of two or more persons living in the same household who are related to one another by blood, marriage or adoption. Also, all related persons living within one household are regarded as one family unit. A household, therefore, may contain either groups of unrelated persons, one or more families, or one person living alone. Thus there are more households than families in the United States, since one-person units are regarded as households, but not families.[6] This is also the reason why the average size of a family is larger than the average size of a household; and why the number of households in the United States has consistently increased faster than the population (parallel to the higher divorce ratios and the increase in one-person households) and why the household's size has gradually decreased—from 4.8 members in 1900[7] to approximately 2.35 members in 1980.

Nevertheless, despite the decrease in household sizes, the relative size of housing units has been increasing: from 250 square feet per person in 1900 to over 870 square feet per person in 1980. This is an increase in excess of 240 percent in square feet per occupant and a 70 percent increase in the average *overall* size of housing units.

One thing that these figures prove, however, is that, although the overall size of dwellings has increased significantly in the last 80 years, their larger spaciousness is also a function of the *smaller* households occupying them. Actually, the number of persons per room has been decreasing to the point where today the United States houses the lowest number of people per room of any nation on Earth: 0.6 person/room.[8]

The size of mobile homes has also increased: from 500 square feet in the early 60's to 720 today.[9] In fact nationwide, the most significant increase in total number of housing units of the last three decades has been provided by this type of dwelling.

Modern housing, on the other hand, has been forced to accommodate an ever-increasing variety of new family structures, all containing increasingly mixed social groupings, and in progressively smaller household clusters. Table 1 illustrates the distribution of household sizes in the United States in 1977.

With both cultural fabrics and socioeconomic structures changing continuously, edifices must adapt accordingly. Consequently, there has been a certain degree of standardization in the programmatic development of shelters based mostly on the flexibility of adaptation to predefined functional characteristics. The successful formulas satisfying these so-called standardized programmatic requirements have yielded the modern prototypes.

Moreover, the development of prototypes illustrates not only social and economic standardizations, but in many cases social fragmentations as well. These trends are easily perceived in housing and residential buildings by the increased development of prototype units for dwellers with similar characteristics. Today this nation is already experiencing "singles only" apartment complexes, senior citizen communities, etc.

There has been much debate about the relative advantages and dis-

Table 1
Household Sizes in 1977

Size	Percent
One person	21
Two people	31
Three people	17
Four people	16
Five or more people	15

SOURCE: Based on information provided by
Heron House Associates (eds.), *The
Book of Numbers* (New York: A &
W Publishers, Inc., 1978), p. 142.

advantages of this trend of building type grouping. With respect to amenities, services, communication and security, these types of structures provide far more benefits for their occupants and their community than does more traditional housing.

ENVIRONMENTS

The shelter inhabitated by the colonists was a rude and austere living and working unit of simple structure and utilitarian nature; aesthetic qualities and appeal were of secondary importance.

During the Industrial Revolution these environments suffered the effects of a new wave of machinism. Shelters were conceived as machines to live in, so that tenements and other strictly functional multifamily units were developed over and over again, ignoring the basic needs of their dwellers or many of the essential characteristics of the human condition. Thus, the relationship between human beings and their places of abode and employment suffered drastic jolts. On the assembly line, the worker could no longer see the finished product of his/her work; children grew up without a clear conception of what their parents did for a living; and impersonal factories, industries and businesses became the producers of goods and wealth and, as such, were looked upon as the real suppliers for the family, while workers became mere production tools.

Under these conditions, spatial environments came to reflect less and less the physical and psychological traits of their dwellers. It was not until the modern architectural movement emerged that new directions and values were explored; but even then, the simplicity and starkness of the new forms were not simply derivatives of functionalism, practicality and the satisfaction of human needs; there were many arbitrary aesthetic choices involved, choices which were defined by the machinism and technological features of the industrial movement which had shaken the world. The power of the factory, the demands of the industrial complex and the wealth of the large business enterprise were the only factors capable of activating or restraining the design trends and products of the emerging architectural movement.

As a result, the control of design features and the creation of these environments moved from the intimacy of the family nucleus to the industrial center and, eventually, to the national corporation. By the 20th century, Americans were caught up in a corporate labyrinth; home, work, health, transportation, communication, goods, all shelter materials and products known to mankind were created, produced and controlled by corporate structures. And thus, work centers and home developments alike fell under the strict control of those efficiently controlled and operated mountains of power and wealth: the corporate structures. These powerful entities directed the eventual consumption and distribution of vast sums of money, the exploitation of unlimited amounts of energy and natural resources, and the employment and labor of thousands of human beings.

The trend has not subsided. Today, despite the fact that the present environmental conditions of most work centers represent a vast improvement over the subhuman conditions of the early years of the Industrial Revolution, edifices still cannot be said to satisfy all the requirements necessary to meet the physical and emotional needs of their dwellers, but rather those of rigid and intricate corporate scenarios.

A great deal of the indifference and hostility expressed by the occupants of modern shelters is due to their inability to understand contemporary structures. One of the most significant differences between modern and traditional buildings is the clarity of their structural systems. Human beings have always felt more at ease when they can comprehend by mere visual inspection the structure that shelters them. In modern architecture, however, most framing members are hidden from the occupants. The log cabin, the Gothic cathedral or the Greek temple were shelters developed around clearly understandable structural assemblages and forms. Today, acoustical tile ceilings, protective skins or decorative panels consistently prevent this essential understanding.

In these incomprehensible modern structures human beings are relegated to evenly distributed work stations of identical nature, which allow for individual means of expression only by the surface application of removable memorabilia.

In many ways, the modern American building is a good example of an unnatural environment. In it, there is virtually nothing that was not purposely designed or conceived in a carefully executed plan. Decoration, form and proportion are strictly controlled and encased into hermetically sealed units dependent on artificial conditions. Within them, human beings feel few climatic changes and experience only the effects of artificially controlled temperature, humidity, lighting or air circulation. Sounds are limited to regulated tones of piped-in music or the white noise of mixed electronic hummings.[10]

In these sense-confining environments, human perception levels are reduced to a minimum with the basic purpose of concentrating the individual's attention on the specific task to be fulfilled at any given moment. Not surprisingly, most individuals rebel consciously and unconsciously in a multitude of ways against these forcibly imposed barriers which define and regulate functions and actions while allowing few means of individual expression. Modern graffiti are more than a blunt protest or a vulgar exercise of freedom of expression; they are the humanization of dehumanized spaces. Most contemporary graffiti are found in the starkness of urban conglomerates for that very reason; in campus areas expressing the intellectual passions of youth on the stark walls of an ''insensible establishment''; in the ghetto, echoing the cries of anger and frustration at the uncaring walls of ''social injustice''; and in a public toilet, liberating repressed intimacies within the confines of a plastic laminate enclosure. Modern graffiti are the reassessment of the human condition on the starkness of modern shelters. Accordingly, the trademark of exposed concrete surfaces has become a four-letter word which, curiously enough, is rarely misspelled.

In many ways, the environment of this nation's modern shelters has been totally reconstructed by objectivity, intention and purpose. And paradoxically, Americans find themselves living within their own creations, but constantly struggling for true freedom of expression, seemingly unable to escape the incarceration brought about by the spatial realization of their own intellectual image.

7
Social Status and Strata

The hierarchies defining social types have been traditionally based on wealth, occupation, educational background, race, age, creed, etc., but in society, individuals maintain multidimensional relationships within these basic categories. Thus an appropriate general definition of social class is the one established by Gerhard E. Lenski, which defines a social class as "an aggregation of persons in society who stand in a similar position with respect to some structured form of power, privilege or prestige."[1] By this definition, individuals may belong to several social classes at once.

In essence, class in America is the respective standing of an individual in a competitive field in which a large number of dimensions and categories define a complex way of life. And thus, depending on the interpolation of these basic dimensions, many different classifications can be established.

WEALTH

Basically, social class as a function of wealth is expressed in the popular format of upper class, middle class and lower class with sub-indexes of high, middle and low in each case.

The types of shelters and the patterns of sectorization of these different classes within urban settings follow similar guidelines in each case. In neighborhood distribution, for example, the sectorization can be easily perceived by simply observing the city map, since almost invariably the sectors of greater wealth are surrounded by areas of lesser wealth which serve as buffer zones to other areas of even lesser wealth, etc., as residential areas grow closer to commercial or industrial sectors. The age of neighborhoods and houses also plays a key role here (with the possible exception of some very exclusive areas), since higher land values increase the land/shelter cost ratio. Barring extremes (ghettos or exclusive "uptown" complexes), most urban costlier land will have the least desirable edifices.

Curiously, a reverse trend (rises in inner-city land values which impose increased housing assessments upon the poorer sectors of the population) has become apparent in certain cities throughout the country. This has caused the urban expansion of a new middle-class-type dwelling occupied mostly by adults with no children to raise who find it appealing and challenging to rebuild older homes and neighborhoods, and who are attracted to the inner city by better transportation costs, tax credits, greater varieties of job opportunities and entertainment centers, etc. Housing shortages, economic declines or suburban decay may further encourage this return to the inner city for some other types of households.

The sectorization of housing patterns within neighborhoods also reflects the socioeconomic standing of the dwellers by defining specific groupings of building types; clusters of shelters with equivalent programmatic layouts; and even lots or street subdivisions with similar features.

OCCUPATION

In this sense the correspondence of social standing to type of shelter is not as easily distinguishable as in the case of wealth.

Certain work groups may try to settle near their work centers and others may not. And to further complicate urban distributions based on occupational patterns, many work centers end up being located (or relocated) according to work force availability.

More meaningful than occupational classifications per se are the spatial requirements that surface through the analysis of the American work force as a whole, and which have had a significant impact on the development of building types. Commercial and industrial buildings offer a fairly good example of spatial differences based on occupational patterns because they are strongly influenced by the operational conditions of their dwellers. There are marked distinctions between an office building and a "professional" office building. Unquestionably, light industrial units, government buildings, banks, hospital/medical complexes, etc. have unique programmatic and aesthetic characteristics. Individually and collectively there is a status difference between those working in professional office complexes and those working in factories, regardless of the physical facilities or income levels attained in each case.

EDUCATION

Although closely related to the occupational patterns, educational factors seem to be losing importance as key social determinants with the passage of time. This trend seems to have been strengthened by the increased mass education movement which followed the Industrial Revolution in America and which gained tremendous momentum after World War II.

Since a college education is much more commonly reached today than a few decades ago, it represents a lesser index of social differentiation. And even if a higher social status is publicly attained by having a college education level over a high school education level, there appears to be a certain "income assumption" associated in each case.

Educational levels bridge wealth and occupational status. However, present trends do not indicate that the gaps in these two particular areas are widening at all. Thus, social standing based on education will most likely continue to decline as better means of learning become available.

STYLES OF LIVING

How people spend their money, where they choose to live, personal appearance, taste, manners, religion, travel, recreation patterns and family background also intervene in the attainment of social status and influence habitats significantly.

Styles of living have common bonds with wealth, occupational and educational factors and even ethnic origins but, in general, people seem to reach higher social status as they embrace recreational or cultural patterns representative of higher social rankings. Playing tennis, for example, is much more fashionable than bowling; and attending symphony orchestra concerts or visiting a museum are looked upon as activities of a higher social standing than attending a country western concert or going to the rodeo.

In the business world, companies also try to maintain (or modify) specific "images" or "characters" aimed at increasing their employees' occupational status, and the social standing of the company itself. As a result of this, the corporate images of Lockheed, Holiday Inn, and Sarah Lee are quite different.

The style of living, as a means of social differentiation, is a highly individualistic trait and will very rarely transcend the limits of a residential lot or a specific commercial property. Nevertheless, certain life-style trends *do* trigger social currents that end up modifying collective living habits. Solar panels, personal computers, home entertainment centers, greenhouses, etc., are all part of movements that reflect in one way or another the importance of life-style factors within the programmatic basis of the human shelter.

AGE GROUP

This is a recent classification factor whose impact in the development of American habitats will most likely become more apparent as the present "baby boom" generation grows older. There are several factors which have contributed to the importance of age as a basic element of social differentiation: in the first place, because people now live longer and traditional age groups have widened, a new socioeconomic class has been created: the "senior citizen."

The youth-oriented North American culture emphasizes traits which characteristically exclude old age groups. Increased mobility and family fragmentations also contribute to leaving older folks behind—individuals with deeper roots and attachments who generally end up populating the decaying areas of neighborhoods and cities—thus contributing to the greater isolation of specific age groups within certain urban sectors or national regions. Also, there are obvious limitations for specific age groups vividly reflected in the type of housing (no stairs, one-story condominiums or low rise buildings with passenger elevators, etc.) and neighborhoods or urban sectors (near communication arteries, health care facilities, etc.) in which these specific age groups tend to dwell.

In its broadest sense, the senior citizens' class may include various categories of wealth, occupation and educational status, but eventually its members can all be clustered within the same general social grouping when it comes to social stratifications on a national scale.

This identification of age groups as a determining factor of social class is most clearly manifested in a recently developed building type: the nursing home.

A brief overview of some characteristics of these buildings will expose some of the basic facts behind age as a social class: In nursing homes there is a marked stress on the lack of physical mobility of the occupants, and their ambulatory freedom is carefully controlled and confined to short walks to and from the sleeping rooms and day room areas. In contrast, the American way of life, in a nation where everything is in constant motion, revolves around a social structure in which mobility is essential. Obviously, what is right at one stage of an individual's age can easily become wrong in the next, but in the nursing home, physical and intellectual immobility are carried to extremes.

In these particular care centers, senior citizens are also rigidly standardized. Activities, interests, behavioral patterns and interactions are all made to follow preconceived notions of the "old age" life-style. Nursing homes also stress artificially controlled environments, social isolationism and class degradation through the mixing of old people with some other types of individuals (mentally retarded, severely handicapped, etc.), who generally represent extremes that suggest the inevitability of the eventual loss of human dignity and self-respect through the aging process.

ETHNIC ORIGINS

This is one extremely controversial area too complex and extensive to be dealt with in detail here. In many cases it has clearly been a factor of social standing. It is an unfortunate fact that often, when a black family moves into a primarily white middle-class neighborhood, housing values in the immediate vicinity decrease. Ultimately, there might be as many economic as social reasons why some of those white families do not welcome their new neighbors, since the effect of racial issues on urban patterns and housing distribution have traditionally been major barriers to the adequate development of urban sectors and even entire cities. If, on the one hand, cultural and ethnic grouping within specific urban areas has some inherent advantages such as allowing better means of cultural expression, preservation of ethnic folklore, and strengthening of traditional values, on the other, this sectorization has also been known to cause group conflicts, urban decay and cultural breakdowns.

Ethnicism, as it affects neighborhoods and communities, is probably one of the most significant areas of influence in the development of urban patterns. If the ethnic groups of the population continue to be primarily concentrated in defined sectors of the urban complex and maintain their present rate of growth into neighboring areas, then their widening suburban boundaries will become the outskirts of ever-growing inner city racial ghettos, densely populated by specific ethnic groupings and troubled by unsettling economies and cultural conflicts.

And it is interesting to note that this is one area where very few specialists have ventured forecasts. Later chapters will deal with the future projections of some of these racial issues and their repercussions on shelters and urban developments.

THE SOCIAL STATUS OF HABITATS

The role played by housing and buildings as representatives of the social standing of their dwellers appears to be closely related to the wealth and occupational dimensions of the previous social stratification categories. Thus the relative standing of specific groupings can be easily defined by merely applying some of the basic evaluative guidelines for habitats and real estate holdings, such as those outlined in Table 2.

The socioeconomic standing of buildings is the resulting average of these particular characteristics once they have been evaluated individually and against one another. But since there are no specific units of measure to formulate quantitative values, the end result can only be a summary of qualitative statements. The same principles and conclusions apply when determining the status of urban sectors in terms of their socioeconomic characteristics.

The general consensus among sociologists indicates that social stratification is much less important today than it used to be, and that it will become increasingly less so in future years. One of the fundamen-

Table 2
Social Status of Dwellings

Neighborhood/ Location	Housing	{ Relative location Lot size/orientation Accessibility/privacy Urban/suburban trends
	Commercial Property	{ Relative location/adequacy Lot size/orientation Accessibility/public exposure Urban/suburban trends
Building Characteristics		{ Quality Size, program, flexibility Collective/private appeal Age, historic value Amenities and surroundings Value and marketability Operation and maintenance Miscellaneous (special purpose, uniqueness, etc.)

tal reasons for this way of thinking is the increased social mobility offered by the meritocratic development of American society.

Traditionally, the social consciousness of "class" has become less significant whenever there have been proportionally fewer and fewer elements at the top, and such appears to be the case here. Affluence, one of the traditional yardsticks for social class differentiation, seems to have lost significance due to the present wealth distribution patterns.[2] Furthermore, a significant mixture of elements is blended within higher income groups, making it difficult to stratify actual "upper classes" since income alone does not seem to satisfy the necessary requirements anymore. Several gifted athletes, for example, have undoubtedly fallen in the category of "millionaires"; yet these individuals, like many other people into entertainment-related fields, have not actually attained the category of upper class. Obviously, wealth alone does not elevate Americans to a higher social stratum anymore.

Strangely enough, although social inequality has always been censored by almost everyone, such is not the case with economic inequality. In several surveys done during the last decade, the general opinion failed to show favorable responses of any significance to the elimination of economic status.[3] Most Americans, it seemed, preferred a certain distinction of economic class as an incentive or reward for efforts and achievements. This is curious in view of the fact that most optimistic overviews of the future build their argument on the attainment of economic equality, classless societies and the total elimination of poverty.

Migratory influxes of one kind or another (to or from specific national regions) will also result in further stratification of housing types, neighborhood conglomerates and social readjustments. Alaska, for example, having experienced a tremendous influx of tradesmen and blue-collar workers in a decade, presently has a population composition very different from that of Florida with its high ratio of elderly Easterners and Latin American immigrants. Similar variations can also be observed within entire regions: there is a widely recognized difference between the South and the Deep South.

Shelters are faithful mirrors of the social rank and standing of their occupants. In trying to step up the social ladder Americans invariably try to improve the appearance, size, quality and style of their habitats. Thus any changes in North American social stratifications would effectively redirect the evolutive patterns of development presently followed by this nation's edifices, communities and urban conglomerates as a whole.

8
Policies, Politics, and Propaganda

In 1670 John Ray noted in his book of *English Proverbs* that "Hell is paved with good intentions,"[1] a common phrase which has been applied more and more frequently to the development of public policies and regulations in the United States. Laws originate as response to readily identified social conflicts or crises and are aimed at their resolution; their implementation, however, does not always measure up to expectations.

Around 500 B.C. Heraclitus maintained that the problem of governmental structures in civilized societies is to combine that degree of liberty without which law is tyranny with that degree of law without which liberty becomes license. That being the case, a study of the implications of the present regulatory system governing the design and execution of buildings in this country reveals one of the most overwhelming tyrannies ever imposed upon society. There is no single factor (with the possible exception of socioeconomic conditions) that affects more drastically the development of modern American shelters than the constraints presently forced upon the building industry by regulations.

The present conditions, however, did not happen suddenly but evolved slowly, starting with the basic needs of the rudimentary and unsafe structures of the colonies and working its way (always based on some kind of "identified need") through a regulatory progression which exploded after the end of World War II and has grown nonstop ever since.

POLICIES

The regulatory intervention in the development of American shelters began in the early colonies with the fire prevention ordinances which banned the use of wood chimneys and thatched roofs, and mandated improvements in old shelters—built as temporary refuges by earlier settlers—being occupied by newcomers. In 1652, in Boston, other less successful regulations dealt with sanitation issues, such as the prohibition of building outhouses within 12 feet of streets or houses, and the keeping of hogs and goats behind fences to prevent their scavenging of outhouses.[2]

Later on, with the subsequent growth of cities, regulations and ordinances began to multiply and deal with some of the social problems and conflicts caused by urban overcrowding, lack of adequate housing, health and sanitation issues, etc. Of the large American cities, the one that developed the most regulations was New York since, among the major urban conglomerates of the second half of the 19th century, it was the one plagued with the most building and housing problems because of its mushrooming urban population.

It was the city of New York that first experienced a significant urban implosion in this nation and as a result, most of the comprehensive local building regulations attempted in this country began there. The Tenement House Act of 1867, for example, regulated ventilation, window sizes and ceiling heights in sleeping rooms, fire escapes, banisters on stairs, distances between buildings and rear setbacks. It also specified water closet or outhouse requirements (one per each 20 occupants) and defined some basic terms employed in its format, such as "tenement" and "cellar," thus setting forth a basis for courtroom arbitration.[3]

The early attempts to codify the development of shelters, however, were only relatively successful. It was not until the Tenement House Act of 1901 was developed that a comprehensive summary of most of the adequate regulations compiled within the preceeding years was established. The new code included definitions of terms which were destined to become significant issues in arbitration procedures: exiting and fire escape regulations; light, ventilation and window requirements; lot

coverages and setback guidelines; and room sizes and public circulation standards. It also covered numerous sanitation requirements. As a
result of this comprehensive Act, the Tenement House Department was
founded, with the primary function of verifying and certifying compliance with the codes sections or issuing penalties and imposing fines
when necessary.[4]

About this time, early building codes began to proliferate throughout the country, and by the end of the 19th century most large American cities had some sort of local building code or housing ordinance
governing housing or building construction.

In 1892 the federal government stepped in by passing a resolution
to investigate slums in cities that had more than 200,000 people. Although it made only a minor contribution, the importance of this federal resolution lies in the fact that the questions over the constitution-
ality of allocating expenditures to enable the federal government to
investigate or interfere with local urban patterns was resolved in favor
of the proponents of the resolution,[5] thus setting a precedent which has
allowed open federal interference with this nation's state and local
building conditions and standards ever since.

The next major federal intervention took place during World War I,
when the government stepped into the construction marketplace by trying
to provide much needed housing for wartime workers. This action gave
rise to the controversy over whether the federal government should en-
gage in the business of providing housing facilities for its citizenship
or not. The controversy ended with the decision that the federal government should provide the housing facilities, but it should also liqui-
date all properties which could not be turned over to other government
agencies once the temporary housing was no longer occupied. In effect, the measure was aimed at preventing the federal government from
becoming a profit-making landlord.[6]

The depression of the 30's brought about more federal intervention
in the construction marketplace with the passage of the Federal Home
Loan Bank, the Homeowner's Loan Act and, finally, the Federal
Housing Administration (FHA) in 1934. Born out of the National
Housing Act of 1934, the purpose of the FHA was *not* to lend money,
but to encourage private lending institutions to do so by *insuring* their
loans. The extent of the Federal Housing Administration's involvement in the housing market, however, went far beyond financial matters from the outset and extended to areas covering regulatory cross-

referencings, inspections, litigations and the eventual development of "minimum" standards for buildings. Later on, with the "GI Bill of Rights" of 1944 the Veterans Administration was created, thus further magnifying the role and control of the federal government over the building industry and its affiliates throughout the nation.

Many other governmental units, legislation and agencies have followed: the Home Loan Bank Board, Federal Mortgage Association, Public Works Administration Acts, Public Housing Acts and Public Housing Administration, United States Housing Authority, Farm Housing Assistance, slum clearances and urban renewal projects, etc. All are ultimately controlled by the Federal Reserve Board and the Treasury Department together with a multitude of legislative bodies and specialized ordinances at all levels and phases of the national regulatory machine such as national building codes, national fire protection codes, electrical and plumbing regulations, and elevator and vertical transportation codes.

These regulations, while having contributed significantly to the improvement of American built environments, have also progressively increased government involvement and control over the construction industry of this nation to a point where today, an average of 50 to 60 different government agencies (or departments within agencies) at the federal, state or local levels can claim jurisdiction on *any* building erected in the United States. From materials, systems and assemblages, testing, standards, inspection and certification to building financing, design and construction procedures, in effect all phases and sectors of the American construction industry are defined, classified and regulated by at least one level of government.

What this overregulated state of affairs does to the building industry in general has been clearly addressed in a survey conducted by the Construction Sciences Research Foundation. A cross section of respondents including academics, building industry consultants, contractors, designers, owners and financial, insurance and legal advisors as well as representatives of most of the professional associations involved in building construction, overwhelmingly affirmed that government interference is the *main factor* affecting the American building industry's capability of producing economical buildings (81 percent majority) in a timely fashion (69 percent majority).[7]

Worse yet than the increased construction cost/time factor caused by excessive regulation are the legal complexities, expenditures and delays caused by regulatory overlapping, as two or more government agencies claim jurisdiction over a specific issue with conflicting requirements. In Wisconsin, the Division of Corrections of the Department of Health and Social Services has often gone on record as recommending against the use of fire sprinklers in detention facilities to avoid false alarms caused by inmates' vandalism. The Building Safety Division of the Department of Industry, Labor and Human Relations, on the other hand, specifically requires the installation of fire sprinkler systems in jails. Furthermore, sentenced prisoners on work-release programs housed in minimum security centers are considered to reside in the "dormitory" classification of the Building Safety Division's code, and must therefore be provided with emergency fire exiting. The Division of Corrections, on the other hand, requires all exit doors to be locked after the curfew hour in these centers; after all, they claim, these individuals are prisoners serving a sentence.

POLITICS

Successfully applied regulations save lives, protect the consumer, guard social rights, establish adequate guidelines for social welfare and prevent social conflicts (or facilitate their just resolution). Unsuccessfully applied regulations, on the other hand, cost lives, rob the consumer, violate social rights, impose inadequate guidelines upon society and provoke social conflicts (or underwrite unjust consequences).

Ironically, even the same regulation applied under similar conditions may end up having opposite effects: fire sprinkler systems, for example, are formidable deterrents to the spread of fire, and in many cases are the best weapon to use against it. However, fire sprinklers have not been proven effective against the release or spread of smoke, and smoke inhalation has long been recognized as the major cause of death in fires. Also, the property losses due to water damage caused by sprinkler systems sometimes exceeds that caused by the fire itself. Nevertheless, fire chiefs throughout the nation generally endorse the widespread use of sprinklers because this type of fire suppression system involves less danger to firefighters. Insurance companies will reward the use of fire sprinklers in buildings with premium reductions, etc.; but these measures are aimed at establishing the capability of salvaging items and avoiding "total losses" that would require insurance companies to pay in full.

Despite their limited effectiveness as lifesaving devices, however, fire sprinklers are still one of the best known ways of combating fires, and that is not an easy task given the state of the art in firefighting: in the last 20 years technology has bombarded human habitats with a tremendous variety of highly flammable materials used in a myriad of unpredictable, unusual or untested ways. Carpeting, for example, which is made out of a wide variety of synthetic fibers (and sometimes highly flammable or toxic gas-producing components), is normally used in flooring. Nowadays, however, the creativity of professional designers has found new uses for carpets and rugs as wall coverings, artwork and even ceiling finishes. Not surprisingly, modern National Fire Codes cover about 16 volumes (over 13,000 pages) prepared by over 2,400 experts sitting on over 175 committees and retail for about $150.00. Meanwhile, fire damages keep on rising.

To overburden an industry as traditional in its basic procedures as building construction with complex regulatory systems is to effectively block most of its viable roads to progress. The tests and approvals presently required by certain regulatory bodies for new materials, for example, should be an area of legitimate concern. Several cost-effective products that have been widely used in Europe for decades have had their use in this country stifled by a myriad of useless bureaucratic prerequisites. And so, while technology and creativity race to develop new dimensions of "progress," public policy and government regulations limp along behind, hopelessly trying to close an ever-widening gap.

Added to these inadequacies are the problems caused by the actual implementation of the regulations themselves. It is noteworthy that in the area of building and construction the proper enforcement of rules and regulations is totally impossible without citizen cooperation. This is especially important where the ordinances governing habitats are concerned. It is very unlikely that building officials will find out about zoning violations—businesses conducted in private residences, etc.—without citizen participation. Yet there seems to be a basic flaw in the implementation of the system: lack of citizen education. Most individuals are completely ignorant of the regulatory requirements governing the development of shelters and—in the majority of cases—cannot identify even the most blatant violations.

Another significant aspect of the implementation of regulatory constraints in the development of edifices is the adequacy of the rule relative to the issues it was designed to govern. Fines imposed for noncompliance with building or zoning ordinances, for example, are passed on to tenants or building occupants in the form of higher rent or "improvement" costs. Others are circumvented through kickbacks, or payment is simply delayed for long periods of time, since often the penalty rate charged for late payment or default is lower than the interest rate paid by financial institutions for short-term deposits.

Conversely, the enforcement of penalties sometimes produces ridiculous contradictions: In a midwestern capital city, the city-owned building occupied by the city government is encroaching approximately eight feet onto public property. By ordinance, it is the responsibility of the building and zoning department to issue notices of encroachment and collect fines (on a rental-basis-type structure) from the violators, thus reimbursing the taxpayers for the use of their property. Obviously, since the city-owned building where the local government is located is ultimately owned by the taxpayers, this situation would require a government body to fine itself and pay the fine with tax money to partially reimburse the taxpayers for the illegal use of their property. In fact, the violation and the fines have been quietly ignored and rightly so, since in this case *not* enforcing the ordinance is the only way to save tax dollars. The legal precedent and implications, on the other hand, are quite dangerous.

By its very nature the government's role in regulating the building industry is likely to act as a formidable factor affecting new systems, procedures and conditions directly responsible for the future development of American habitats.

PROPAGANDA

One of the most paradoxical problems presently facing the evolution of buildings in the United States is the way in which most edifices are misrepresented in the public media.

This misrepresentation covers issues which should not be underrated no matter how minor they may seem, since some can have serious consequences: stunt scenes on roofs can lead people to believe that jumping from roof to roof or running over steeply sloped planes is an easy task, when the fact is that walking on *any* roof which exceeds a 3.5:12 pitch is highly dangerous, no matter what the roof surface may be.

Window ledges and metal gutters shown on television and in motion picture scenes appear far stronger than they really are: *there are no strength requirements for live load support for gutters or decorative ledges in any American building code whatsoever*. Actually, the relative resistance of those construction items depends entirely on the particular details devised for their attachment. Action scenes in movies and television shows are done by professionals and take into account a myriad of safety measures not present in average building structures. Yet there are no clear warnings of this anywhere. One may easily be led to believe that the building features and strengths displayed in most action scenes are typical of construction standards everywhere. They are not.

Conversely, doors (and exterior doors specifically) are a lot stronger than they appear to be on the screen. Trying to kick a door open, or throwing one's weight against it, is much more likely to result in a broken leg or a dislocated shoulder than in a real-life replay of a heroic scene by Kojak. It is also very unlikely that anyone weighing over 20 or 30 pounds can get away with swinging from a ceiling lamp without breaking the ceiling plaster board holding it. Nowadays it is very rare to find hanging lamps fastened directly to structural members. What Douglas Fairbanks, Jr., or Zorro may have seemed to achieve so easily may cost an average human being a seriously injured back or his/her life.

The effects of the misrepresentation of the human shelter, however, go beyond a few broken bones or the danger posed to a limited number of lives. The most serious consequences involve the home of man conceptually, since by misrepresenting some of the basic characteristics of habitats the public communication networks end up misrepresenting society as well.

Until recently, on any given Sunday, one could watch three consecutive television programs on one of the major networks depicting three different housing categories: the urban town house of a factory worker and all of his family, a middle-class apartment where a divorced mother struggled to raise a family on a day-at-a-time basis, and the luxury apartment of an east-side Manhattan complex where a wealthy minority couple lived. If one bothered to count the paces it took the average person to cross the living/dining area in each setting, the size of the three dwelling units turned out to be the same (not surprisingly, since the actual sizes were dictated by the size of similar studio sets); and ironically, the kitchen of the blue-collar family ended up being approximately twice the size of the kitchen in the luxury apartment. In each case, by the way, the size suggested to the viewer was approximately 32 feet by 16 feet for a living/dining area. Very few middle-class homes and certainly very few typical apartments have living/dining areas that large.

The results are obvious: the human interactions seen on the television set do not correspond spatially to those which take place in real life. Blue-collar town houses similar to the one shown on the television show are typically much more cramped than those depicted on the program, and the personal spaces experienced by families in real life are much smaller. For the blue-collar worker or the unmarried working mother in similar economic circumstances the comparison between his/her shelter and the one shown on television—whether conscious or unconscious—*has* to be frustrating.

People are never crowded on television by actual space limitations unless it is by a specific dramatic requirement; actors are rarely forced to interact within the confines of a three-foot-wide hallway. Some of the most essential interactive characteristics of human beings and their shelters are entirely omitted in the portrayal of homes on movie or television sets. Furniture, for example, always seems to float—due to obvious performance requirements—because there is always one wall missing: the blind partition against which one commonly finds the most significant furnishings, wall decors and items of personal expression of any household.

Soap operas have standardized the settings for coffee breaks, office decors, furniture styles and interior design features according to profession, social status and personality.

Almost invariably the exterior and interior representations of shelters are in conflict with one another: a person is shown arriving at his/her home during the late afternoon hours and entering a regular size house casting a shadow to the observer's left and having a picture window in front. This is, obviously, a house facing North. Yet once inside, the picture window turns out to be about twice the size originally perceived and the sun is shining directly *through* it. Suddenly, by a simple door slam and the magic of celluloid, the house has been turned around approximately 90°.

Urban houses also have a tendency to grow tremendously once the cameras enter them. Since exteriors are factual shots and interiors are often done on television or motion picture sets, the sizes rarely correspond. A clear example of these altered facts is the continuous exposure of large bedrooms, wide halls, etc., in old-style urban houses. Except for a few rooms in castles and plush estates, most old urban homes had—at best—average size bedrooms compared to those in modern middle-class housing. By the present spatial standards, most old urban houses were large structures filled with rooms of small to average dimensions; yet this characteristic is seldom suggested in motion pictures or television programs where old urban houses consistently have an air of bigness and spaciousness very uncharacteristic of their times. The typical bedroom shown in many cases—whether master bedroom, guest room or nursery—appears to be at least 20 feet by 25 feet.

To this, one must add the angles of focus and corrective lenses used to make spaces appear in perspectives which rarely correspond to fact. Architectural and interior design magazines are specialists in this kind of distortion. Interior (and to a great extent, exterior) photography has become such a specialized art that if it were not for the repeated ferns, spider plants and philodendrons which appear in the photos of several different rooms, one would be led to believe that almost all the spaces of the shelters shown in these publications were epitomes of perfection.

Wide-angle and other perspective-distorting lenses can work miracles with a mere eight-foot by ten-foot room. However, television documentaries that show live action indoor scenes offer a much different picture, since the proper viewing angles are inevitably so difficult to achieve that the camera ends up filming nothing but "close-ups" or "waistline takes."

Spatial misrepresentations about the size and proportion of shelters also distort the human conception of space/time and man/space relationships. Filmed sequences, for example, compress the time it takes a newcomer to find a specific suite in a large office complex to fractions of a second—obviously for time/cost and attention span reasons—but with a totally unrealistic representation of what space/time factors in large office complexes are truly about. On the other hand, anyone visiting the White House for the first time will probably be surprised at how small some of the rooms actually are when compared to modern spatial layouts, since most preconceptions lead everyone to believe that buildings of this time and category had oversized proportions, when the fact is that—precisely because of its age—the human scale is better observed in the White House than in many modern affluent estates.

American public information systems have often been criticized for their inaccurate representation or biased interpretation of social, economic and political factors; but if adultery, incest, poverty, welfare, crime and corruption are all issues which may affect this nation's citizenship at one time or another, the inaccurate representation of shelters is an issue that affects each individual in society *directly*.

The "baby boom" generation of the postwar years has been the first generation of Americans to grow up with television. It will be in this decade, as this generation reaches the homeowner's age bracket, that the primary effects of these circumstances will be felt in the future development of housing and buildings. This is a significant fact and, consequently, will have a significant impact. Crime, divorce, famine or affluence are isolated issues that the majority of American citizens do not have to deal with twenty-four hours a day, every day of their lives. The human shelter, however, *is* such an issue.

9
Buildings and Construction

In all societies the three elements that absorb the highest proportion of investment are the supply of raw materials, the supply of food and the supply of shelter. The provision of housing and buildings and their associated infrastructures (streets, utilities, construction materials, advertisements, etc.) is, accordingly, one of the basic pillars of the socioeconomic structure of any nation. In a sense, the construction industry is more similar to agriculture than to a manufacturing industry, since building construction does not fabricate "needs" for the consumers of its products, but merely satisfies existing market demands and thus is not an exploiter of consumerism.

Construction in the United States has been defined as a risky and highly fragmented industry in which none of the constituents occupies a dominant position. It is highly competitive and passive in nature—that is, it responds to natural demands on a short-term basis—and is inadequately managed, with very little invested for research on itself and its future.[1] In effect, the so-called American building industry is not an industry at all, but a conglomerate of enterprises which come together for the specific realization of building projects only to disperse again once the project is completed.

The construction industry employs approximately 5.5 percent of the work force in the U.S.A.; but this is a misleading figure, since it refers strictly to those directly involved in building construction and does not include supporting tasks such as clerical help, retail businesses, mining operations, product manufacturers and labor, government officials, sales and advertising, real estate, banking and insurance, public utilities, etc. In practice, the ramifications of the construction industry in this nation reach every group of the economic spectrum.

The basic factors which control the performance of the building industry are: resources, economy, technology and society.

Here is a brief synopsis of the correspondences between each of these factors and the development of habitats in America.

RESOURCES

Land and the Environment

Contrary to what overpopulation statistics would have everyone believe, the United States ranks 4th in the world in fewest persons per square mile of national park land with an average of 1.0 inhabitants per square mile, behind Canada (0.26), Australia (0.28) and Norway (0.30).[2] This is all the more laudable when one compares balance of population densities (inhabitants per square mile) of these four nations: Canada (6.7), Australia (5.0), Norway (34.3) and the United States (62.3). America also ranks 4th in homeowners as a percentage of the total population (64.6 percent), behind Ireland (70.8 percent), Australia (67.3 percent) and Canada (65.4 percent).[3]

When looked upon in context, this nation's preoccupation with adequate shelter for its people has been extremely well managed. Despite the negative aspects of suburban and urban sprawls, housing developments occupy on the average as much as 60 to 70 percent of the land area of American cities and represent in excess of 32 percent of the overall construction dollar.[4]

Investments in shelter development are significantly affected by the supply of money. The cash used to pay for the actual construction of housing and utilities is approximately one-third of the cost of the housing; the rest goes to pay for financing, land and taxes. One of the most serious problems regarding these investments lies in the increasingly high costs of land development not because of land shortages but because the preparation of acreage for suburban expansion is a very complex and costly process.

The high price of urban land is a direct reflection of the large public and private expenditures necessary for its development; communication arteries, utilities and zoning changes basically depend on the willingness of governmental bodies to undertake development tasks that usually end up burdening small communities—which consistently oppose urban growth—with higher taxes. Moreover, the unavailability of land to be converted for building development has many economic advantages for the residents of an area, since the shortage of habitable land typically drives the value of already developed land upwards.

Land for shelter development is an available resource, but it is being driven to scarcity by the influence of special interests, public policies and private speculation. Hence, the relative percentage of land cost versus building cost as percentages of the overall project's cost has climbed progressively since the end of World War II, with the indexes showing a rate of increase from 20 percent of total cost for land and 80 percent for buildings in 1950, to about 30 percent for land and 70 percent for buildings in 1970—a relative increase of 50 percent in land value in just two decades.[5]

This is a trend that continues to be on the rise: in some areas housing presently evaluated at around $80,000 for the dwelling alone may command total real estate value of $120,000 to $150,000 due to the inflated cost of land. Several national forecasts, for example, have the land/building ratio reaching figures as high as 45 to 55 percent by the turn of the century.[6]

The lack of available land and its burdens on real estate will constitute one of the most powerful determinants in the future evolution of American habitats, but as a resource, the problems of land development seem to lie more in its processes of transformation than in its overall availability.

Such is not the case with environmental issues. In this respect, the damages can be irreversible. Continued increases in the levels of rainfall acidity, air and water pollutants, etc. may eventually redefine completely the materials and coatings presently employed in the skins of shelters.

Materials

Most building products derive from the most prevalent materials on Earth, and except for the danger of uncontrolled environmental damage, there is no reason to believe that this trend will change in the near future. Despite the fact that some construction materials may appear to be headed for scarcity, it is very unlikely that the development of shelters will ever be inhibited by the unavailability of any particular product, since the construction of shelters cannot be developed around items of precarious supply and almost always can be easily redirected to incorporate more abundant substitutes.

Because of their abundance, as well as their ease of replenishment or recycling, wood, concrete, plastics, steel, aluminum and glass have consistently been heavy favorites. Wood has certain properties which make it an ideal building material: a simple 12'' × 2'' × 2'' cross section can resist 40,000 pounds of end pressure. Wood is not damaged by severe temperature or climatic changes; tests have shown that wood becomes stronger at 300° F than it is at 70° F and, if chemically treated, it will not burn at all.[7]

Concrete is another plentiful material which may remain a favorite in building construction for quite some time. The Romans were the first to use what today is known as concrete (from the Latin *concretus*, "growing together"), but with the fall of Rome the use of concrete construction was lost until the 18th century. It was not until the development of portland cement that concrete became a popular building construction material once more.

Portland cement is a relatively recent discovery. It was developed by Joseph Aspidin in 1810 when he was searching for a water-resistant mortar with which to build the lighthouse at Eddystone in England. He called it "Portland cement," supposedly because the material was similar in color to the stones found on the island of Portland in the South of England.[8] Steel-reinforced concrete was developed shortly afterwards by Joseph Monier and first used in the United States to build a house in Port Chester, New York, in 1857.[9] Variations in use and a gamut of admixtures have progressively improved the characteristics and applications of concrete and concrete reinforcement systems, thus popularizing its use throughout the world.

Steel, aluminum, plastics, paper derivatives and other materials are also among the most abundant of Earth's resources. The problem, therefore, is not one of the availability of the basic raw materials but of shortages of "satellite" products which are employed in the production or processing of other basic resources: manganese, bauxite, chromium, cobalt, zinc, tungsten, gypsum, copper, lead, etc. In this sense, it would seem that in the near future the advances in building construction will most likely come more through breakthroughs in the manufacturing processes of certain products rather than through the discovery or implementation of new and revolutionary construction materials, with the possible exception of plastics and synthetic fibers or some highly specialized metal alloys.

Labor

Building construction has always been a labor-intensive industry because almost all buildings have traditionally been erected on site. And the characteristics of the construction work force are rather unique since it must admit a high percentage of unqualified laborers due to the inherent expanding and contracting demands of the marketplace.

The manpower employed in construction must also be highly mobile, continuously moving from building site to building site within specific regions or even from region to region within the country. Even contemporary prefabrication processes serve only to move the laborer from the construction site to the manufacturing plant; but the general tradesmanship, qualification and production output remain quite similar. Thus the only significant achievement of most prefabrication processes employed by contemporary construction technologies are essentially higher degrees of quality control.

When unionized, labor also affects the structure of the construction industry drastically through restrictions on equipment, methods of construction and productivity indexes. Examples of this are plumbing unions, which have consistently inhibited certain prefabrication procedures by insisting on plumbing being built on site; painters who have been known to limit the size of brushes and the use of spray guns; and masons who have limited the number of bricks placed per man per day.[10]

THE ECONOMIC FACTOR

In the broadest sense, an economic analysis of building construction in America includes a multitude of facets ranging from macroeconomic studies (construction dollars as percentage of gross national product, government outlays for construction, etc.) to a myriad of micro-economic analyses (such as the correspondences of household units, housing types, economic bases, and others). These studies have traditionally provided the basis for all building construction forecasts in the United States.

There have also been attempts to develop macroeconomic theories of cyclical evolution, which theories, although not widely proven or accurate, are worth mentioning, since they describe in general terms the interaction between economic fluctuations and the development of habitats.

A cyclical sequence can be defined as follows:[11]

1. Incomes rising period. As incomes rise and spatial capacity is saturated, rents increase.

2. As rent increases, building owners realize higher profits. New construction becomes more profitable and investments in new construction begin.

3. New construction exceeds market demands. Building development revenues decline because of increased availability and declining market demands.

4. Credit tightens and construction activity declines. Market is saturated. Incomes decrease.

5. Production and demand begin to increase labor activity once more. Occupancy tightens within available shelters. Incomes begin to rise.

There have been no properly identified long-term cycles in construction that have followed any specific pattern to date. Most short-term cycles (seven clearly identified since World War II) have been caused mainly by fluctuations in the availability of construction financing mostly due to governmental monetary policies.

Overall, investment in the development of shelters comes from the following financial institutions: savings and loan associations (23 percent), commercial banks (23 percent), life insurance companies (24 percent), mutual savings (10 percent) and others—such as pension funds, bonds, etc.—(20 percent).[12] It is generally distributed as shown in Figure 2.

The actual cost of shelters, on the other hand, offers a microeconomic picture which describes quite clearly its significant dependency on three basic resources: materials, labor and land. Table 3 illustrates the typical breakdown of relative 1980 housing costs of components.

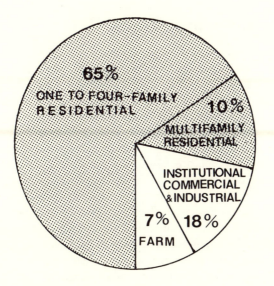

Figure 2
Shelter's Investment Distribution

SOURCE: Based on data provided by Elena Zucker (ed.), "CII Loans," *Building Design and Construction* (January 1980): 56.

Table 3
1980 Housing Components Percentage Costs

Component	Cost % of Multiple Components		Prorated Cost % of Basic Components	
	Single Family	Multi-family	Single Family	Multi-family
Materials	32	33	44	47
Labor	24	23	33	32
Land	16	14	23	20
Profit	9	10		
Overhead	8	6		
Marketing & Sales	5	4		
Advertisement	1	3		
Others	5	7		

SOURCE: Based on data provided by Richard J. McNeilly (ed.), "1980 Profile: The Builders of America," *Professional Builder* (July 1980): 76.

THE TECHNOLOGICAL FACTOR

From a technological standpoint, innovations in the construction of shelters occur either through the direct incorporation of new products such as power tools and innovative materials—over some of which government regulations exercise a significant amount of control—or through the application of new technoscientific *concepts* or principles. The latter are comparatively major types of contributions since they include new construction systems and procedures (prefabrication, tilt-up and stack-up systems, concrete slip forms, etc.).

The building industry has felt the impact of 20th-century machinism in a multitude of ways: in the transportation and manufacturing of materials, the use of heavy equipment, serialization and standardization, etc. However, it is noteworthy that despite this, most tools and equipment employed today in the realization of the American shelter (from cranes to nail guns) are aimed at improving *traditional* construction procedures; backhoes may perform better and faster than hand excavating and nail guns may drive fasteners at strengths and speeds no human arm could expect to achieve, but excavating, sawing, nailing and drilling are still basic processes in the development of shelters.

In addition to "modernized" versions of traditional tools, there are proportionally very few specialized pieces of equipment developed specifically for construction as a result of new construction systems. Rather, most new tools are merely suited to new methods of packaging or applying traditional products: putty guns, drywall tape rollers and metal roofing seamers, to name a few. To illustrate this limited technological progress in building construction, one only needs to point out that within the last decade the tool which has had the most overwhelming impact on the processes of shelters design and construction has been the pocket calculator, since it is the only tool that has completely replaced its predecessor: the slide rule.

SOCIETY

In addition to the socioeconomic interactions, cultural traits and regulatory matters, the impact of social issues on habitats extends also to areas which affect building construction more subtly, such as variations in the basic characteristics of the industry work force, special construction requirements and real estate constraints.

The imposition of women and minorities in construction has contributed significantly to the increase in building costs. Accessibility requirements for the physically disabled have also involved enormous capital expenditures throughout the building industry: recent studies have shown that public building projects cost approximately 10 to 15 percent more than private construction precisely because of the extensive paperwork and affirmative action requirements involved in them. Proponents of the barrier-free environment requirements in construction claim that the overall cost of providing buildings with these features represents a very small percentage of the overall national expenditure in construction. The statement is misleading, however, and it certainly does not mean that compliance with barrier-free requirements in *all* buildings is a relatively inexpensive proposition.

Land development is another area seriously affected by social issues; as recently as 1980, the United States Justice Department was reportedly weighing whether zoning laws that establish minimum lot sizes are discriminatory, since they increase housing costs and thus make those lots unaffordable to poor minorities and other dispossessed citizens. Meanwhile private neighborhoods and clubs, self-contained mini-communities, etc. flourish, as social behaviorism seems to defy regulatory intentions in principle and in practice.

NEW PROMISES, NEW WORRIES

Technology is the practical application of scientific discoveries. Any new discovery alters the flow of events by introducing new facts and by imposing new concepts upon society.

In essence, technology is a means of creating needs. In Daniel J. Boorstin's words, "Technology is a way of multiplying the unnecessary. And advertising is a way of persuading us that we didn't know what we needed. Working together, technology and advertising create progress by developing the need for the unnecessary." [13]

Traditionally, the essential structure of North American society has been redefined by technology, and in all likelihood, the future wants of its citizens will be mostly defined by technological developments.

This brings up an interesting question about the future development of American habitats, and that is whether the trend will be that of edifices adapting to man or that of man adapting to edifices. On the one hand, new possibilities seem to favor the adaptation of the shelter to man as a viable trend: portable structures, prefabricated units, geodesic domes, etc. are shelters which, once designed, built, or modified to suit specific users, can *relocate* with their occupants. Moreover, new materials (plastics, paper products, wood derivatives, and lightweight materials) also seem to reinforce this trend.

On the other hand, there are factors which would not facilitate the adaptation of shelters to man, but would rather have the opposite effect: higher costs of living, economic changes and population shifts, increasingly complex real estate transactions, stagnating land development and the unavailability of public utilities among others.

As far as the near future is concerned, the most likely possibility is a mixture of the two tendencies: the reciprocal adaptation of man *and* shelter. However, the long-term prospects are quite a different matter. There the trends are based on patterns that extend far beyond specific socioeconomic climates.

One significant characteristic of technological developments is their lack of justification. In the broadest context, technoscientific revolutions have never had a purpose. The Industrial Revolution of the mid–19th century *caused* drastic social and economic changes, but it did not develop as a result of any recognized social or economic needs. Man did not land on the moon as the result of a practical necessity; nor were radio, television, telephones or automobiles created out of the "need" for improved communication.

The implications of technological advances can only be fully understood when seen in retrospect. Neither the need for technoscientific improvements nor their effects can be recognized from within the specific time span in which they happen.

In this light, exaggerated concerns or expectations about scientific discoveries are not justifiable. It is in the *known* technoscientific world that factual grounds for hope or concern are to be found. Paint and varnish removers, insecticides and many new plastics—all products recently incorporated into American homes and buildings—are widely recognized allergenic or poisonous agents. Only recently, the cause for Kawasaki's syndrome—an illness prevalent in children which leads to heart problems and aneurysms—was linked to the method and substances used in rug shampooing.

The danger posed by the introduction of new substances and chemicals in the American built environment is not an issue to be regarded lightly since the hazards, like the need for these materials, are strictly man-made. And there lies the challenge: in dealing successfully with the problems caused by the constant incorporation of manufactured needs into the modern way of life. However, the changes necessary to meet this challenge have its drawbacks, since they are based on the assumption that the interaction between new technologies and social transformation is invariably conducive to social welfare, whereas it could also be viewed as causing the servitude of man to machine. Western man once used the core of his energy to build churches and cathedrals; today he uses the totality of his resources to build laboratories and nuclear reactors.

As a result of its explosive machinism, North American society has experienced continuous drastic change since its inception. Overall, one could say that, since the 1920's, just as the technological developments have caused drastic changes in this nation's social structures, they also have restructured the general layouts, spatial distribution, functional relationships and stylistic requirements of its habitats.

Despite the assertion that the shelter of man is basically functional (since its high cost limits its adornment with the superfluous), a gamut of unnecessary built-in items have become common features in modern shelters simply because they are considered "convenient" or "appealing": door closers, curtain walls, reflective or tinted glass (one can achieve the same energy efficiency by *reducing* the glass area in a wall), wallpaper coverings, trash compactors, intercom systems, multiple telephone outlets, artificial climate and garbage disposals, just to mention a few.

The increasing use of home computers promises to revolutionize the average American shelter. Yet, the presence of a computer in every home can also be a cause of concern. Like the telephone, a home computer is a window open to external intrusions; but unlike the telephone, it permits unannounced intrusions. Files on tax returns, bank accounts, medical files and transcripts, credit ratings, legal matters, social security information, personnel files and telephone records could all be combined to form a dossier on an *individual basis* accessible to anyone who successfully breaks the computer code (a feat which is becoming increasingly common). The probability that personal mail will eventually be handled through electronic networks presumes another giant step that involves some very real concerns. There is more to home computers than accounting records, telephone answering services or video games.

Unquestionably, the soaring technological advances of the last few decades are at the core of the overwhelming cultural developments experienced by this nation. However, not all new technology is beneficial, adequate or infallible.

For example, although the use of computerized structural design systems and the safety margins presently employed in modern building construction represent significant improvements over those found in earlier years, modern structures still suffer from numerous hazards and inadequacies. Recently, serious structural failures have injured or killed a large number of building occupants throughout this nation. The collapse of the Hyatt Regency balconies in July of 1981 was not a unique case; within the last three years alone failures, collapses or remedial construction have also occurred at such places as the roof of the Hartford Civic Center, the Long Island University Dome, the roof of the Rosemont Horizon Arena near Chicago, the Kemper Arena roof in Kansas City, the University of Florida's O'Connell Center, the Civic Center in Worcester, Massachusetts, and the Brendan Byrne Arena in New Jersey. Every year, an average of 145 bridges collapse in the United States.[14] Anyone believing that modern construction is free from failure or that present computerized structural designs employ infallible computations or safety factors is seriously mistaken.

On an urban scale, such areas as transportation, urban decay and crime are also subject to technological promise and technological concern. New distant subdivisions increase private automobile usage as a means of transportation and it is noteworthy that the most common criticism of unchecked urban growth is its imposition of a feature which is barely one hundred years old: asphalt pavement, the first of which was completed in Newark, New Jersey, as recently as 1870.

The proliferation of the private automobile as an urban transportation unit is a modern dependency of questionable value. The American family car represents, on the average, 2,000 pounds of steel powered by a 250-horsepower engine which transports between 1.1 and 1.5 persons per trip;[15] and these trips seldom involve emergencies, but are merely short unplanned or spontaneous runs to schools, churches, social gatherings and shopping areas.

The widespread use of this inefficient form of transportation has actually made traffic engineering a profession concerned more with the resolution of problems than with the planning of city traffic patterns. Thus the remedial solutions are merely short-term answers to deeply ingrained urbanistic transportation problems. Consider this: automobiles on any specific street must yield to cross traffic by means of stoplights (so far the most effective way of controlling cross traffic without the enormous expense and consequential problems of building overpasses). On equal density arteries, the complete stoplight cycle averages one minute; that is, cars traveling in each direction stop or move for thirty seconds at a time. The overall traffic effect of these light-change cycles is a 40 percent reduction in the total number of cars which could circulate uninterruptedly through any specific intersection.[16]

The conversion of streets to one-way traffic during peak hours, detours to larger arteries and a multiplicity of auxiliary systems are often employed with greater or lesser degrees of success, but the basic problems persist. In the long run, urban traffic problems do not respond to the use of palliatives. Perhaps one of the greatest paradoxes of modern times is the enormous amount of attention devoted to facilitating automobile travel as a means of improving congestion, when it is actually the least efficient form of urban transportation. In a similar fashion, the bus replaced the electric train even though it is a much less efficient form of rapid mass transit. A typical street with average intersections will accommodate approximately 750 cars per lane, per hour, or 180 buses or 150 street cars; but since cars carry an average of 1.5 persons, buses 40 and street cars 50, a single lane would handle 1,200 people riding cars, 7,200 riding buses or 7,500 riding street cars. Also, since the buses and street cars could carry standing passengers, the total capacity could be further increased to 9,000 people per lane per hour for buses and 13,500 people for street cars.[17] This efficiency ratio translates into a tenfold increase from automobile usage.

Technology may promise new and improved vehicles—perhaps lighter and truly energy-efficient ones—which could run on water, batteries or even solar powered cells, and be noiseless, pollution-free and affordable by nearly everyone. This promise is extremely appealing to the average urbanite, but constitutes a nightmare for the traffic engineer and some of his fellow victims of the effects of automotive expansion. Recent statistics show that the United States leads the world in deaths from automobile accidents, injuries from automobile accidents, and automobile accidents per 1,000 habitants, while allowing the *lowest* maximum speed on motor ways and secondary roads of most nations on Earth.[18] There is good reason to fear this nation's overcoming the automotive fuel crisis.

AFFLUENCE AND SCARCITY

The 1980 census recorded 88,394,574 housing units—single and multifamily—in the United States.[19] If one multiplies this total by a modest $70,000 cost per dwelling unit, and to that figure one adds the value of other commercial, industrial and public housing, the resulting figure would be staggering and obsolete by the time it was calculated, since the cost of shelters has been rising at ever-increasing rates in recent years. Despite a depressed housing market, the 1982 average cost for detached single family dwellings of $92,300, for example, meant an 80 percent increase from 1975 alone.

It is precisely because of the enormous investment that habitats represent in this nation that their development cannot support revolutions of the kind predicted by many forecasters. The 42 million plus detached single family dwellings of the United States alone are an investment well in excess of 2.5 trillion dollars.

The depicted images of cities of the future with skyscraping towers sustaining bubbles at the top may be entertaining and imaginative, but they are inconceivable according to all present patterns of spatial development. A hollow column—of *any* material—50 stories high encasing a vertical transportation system—*any* vertical transportation system—and supporting a habitable structure on top—*any* habitable structure—is definitely an extremely inefficient way of developing a building, and is only justifiable in very specific cases (such as lighthouses) due to some unique function.

Because of the higher price of shelters, cities have traditionally embraced old and new alike. Urban conglomerates are, in effect, *dated* by their buildings. The economic risks involved by the introduction of drastic innovations in built environments are far too great to support spatial revolutions. The inherent permanence and lasting qualities of edifices also help increase their economic magnitude: if a dweller knows that he or she will use a product for a longer period of time, it makes sense to buy a more elaborate item—of more lasting quality—since the extra cost can be spread over a long period of time. Accordingly, even if technology were to reduce materials and construction costs, the lower cost indexes would not bring equivalent reductions in the total cost of shelters; they would simply result in larger, more complex and more elaborate buildings or buildings of a higher quality.

Continued affluence in America will undoubtedly influence the development of its habitats, but it is very unlikely that any sudden revolution will totally redefine buildings or cities throughout the nation. Even if some unforeseen economic phenomena were to facilitate variations in shelters, most of the basic factors acting upon the development of built environments would have to be modified prior to adopting any drastic changes. Underground buildings, for example, may seem very appealing to many, given the present state of the art regarding environmental and energy conservation issues. Regulatory matters, however, as well as the inherent dangers dealing with fires and fire prevention, toxic and allergenic substances, etc., represent formidable barriers to the generalization of this type of shelter evolution.

The impact of negative environmental developments and resources availability—land, materials, labor, etc.—are also aspects which have been consistently misinterpreted. No one will ever wake up to a world without chromium where nothing shines or a nation without parks, trees or birds. As in the case of other items whose availability has dwindled, the first effects of their scarcity will translate into economic questions. With the known supplies of gypsum and other basic construction materials dwindling on a worldwide basis, how will the availability of resources affect the affordability of adequate shelters in the future? What new material substitutes could be developed? And at what price?

The evolution of this nation's habitats has been closely linked to its socioeconomic and technoscientific cycles. When machinism ruled during the Industrial Revolution, the American shelter was redefined as a "machine to live in," conceived as such, built by machines and filled with mechanical items. Today, it is again being redefined as an "electronic cottage" which will eventually be designed, built and maintained by computers. The similarities between both attitudes are staggering. In this sense, it appears as if although shelters are eventually conceived with man as their unit of measure and functional model, man's own image becomes a function of technological postulates. And thus the evolution of modern shelters has become a consequential phenomenon that employs the technological image of an era as yardstick and its level of affluence as mortar.

TODAY AND TOMORROW

Experiments conducted at Princeton aimed at altering human perception of time indicated that when the concept of the future was eliminated in subjects, they became euphoric and anxiety-free, and described their feelings as semimystical in nature. They also experienced loss of identity and ambition, and, overall, became less motivated. Inversely, with an expanded conceptualization of the future they became happy, fulfilled, calm, and, in general, lost their fear of death.[20]

The concept of time is a key determinant of human behavior. Today and tomorrow are so interrelated that neither can be conceived without the other. The same experiment, by altering time perception, produced immobility and feelings of depression, hostility, loneliness and withdrawal in subjects whose concept of the present was eliminated; while exuberance, fascination and slowness of time perception resulted when the present was expanded.[21]

Ventures into the world of possibilities and speculations—no matter how accurate or far-fetched they may seem—are essential for anyone who would seriously plan for the future and, in so doing, structure it. After all, the foundations of tomorrow are being built today.

So as the nation navigates the uncharted waters of scientific discoveries propelled by its consumerism, steered by social changes and guided by its economic compass, it is time now to look ahead and scan what may lie on the horizon, and beyond.

PART 2
PROSPECTS

10
Time and Time Machines

The human perception of reality is a direct function of the human sense of time. Without the sequential awareness of change, mankind's role on this planet would have been no different from that of any other animal species. Time is strictly a human concept and one of the most significant characteristics that separate human beings from the rest of the animal kingdom.

The concept of time has interested man since very early in history. As early as 450 B.C., Zeno of Elea began to deal with the idea from two different—yet complementary—points of view: time as a continuous flow and time as a sequence of intervals. The distinction is illustrated in the paradox of Achilles and the tortoise where it is maintained that Achilles—although a much faster runner than the tortoise—could never catch the animal in a race if the tortoise were given an initial advantage (no matter how small), since with both runners starting at the same time, Achilles would have to reach midpoint between the distance that separated them; but by then, the tortoise (also moving) would have advanced some, so that upon reaching the first midpoint Achilles would have to set out to reach the midpoint of the new distance, and so on. Thus it was concluded that Achilles could get infinitely close to the animal, but never actually catch up with it.

The fallacy in the paradox is that in it, time is only cut into ever smaller segments and the concept of time as a continual flow of events is totally disregarded. Obviously, no one can move ahead in a world where time is continuously reduced into smaller sequential fragments. This statement, however, should not be lightly passed over since modern society, due to its complexity, is actually segmenting time spans in a multitude of subtle ways. Ironically, the effects of this can be easily observed precisely in situations in which man is released from the confines of segmented time spans, such as the disorientation experienced by most new retirees when they find themselves free of rigorous schedules. The compartmentalization of functions and the increasing emphasis on predefined activity/time span segments in modern society are nothing but definitions of "midpoints" in a race where human beings set out to reach goals established by a myriad of stereotyped social standards.

Man has always been disturbed by his mortality. The struggle to unravel time's secret has preoccupied the mind of every great thinker in history. From Aristotle and Plato to Augustine and Kant, the effort dedicated to the understanding of time by the most profound and brilliant minds among men has been enormous. Underlying their diverse conclusions, however, is one common belief: time is the principle of order in the human understanding of the world as it is perceived within a framework of sequential occurrences. Thus time and order are essentially related, and so are time and human perception. In fact, all time known to man is experiential time: what is memory as past, what is perception as present and what is expectation as future.

The human understanding of time, however, is not as clear as man would like it to be. Time as an unbound measure of order lacks the parameters that would make it readily comprehensible by the human intellect. It is so closely linked to infinity that its full apprehension seems forbidden to mankind. Even in quantifiable form, the true significance of time lies beyond the limits of intellectual grasp. Although it may be easy to state that Earth is believed to be about 4.5 billion years old, how can anyone comprehend a figure as immense as that? The attempt to do so has led to the use of a technique known as "time compression," which is a process much like bridging the gap to an incomprehensible concept by developing an understandable scaled model of the sequential passage of time. A time compression model of Earth's age, for example, would equate the entire development of the planet (the full 4.5 billion years) to, say, one calendar year. It is then a lot easier to establish comparisons since at that scale one would realize that dinosaurs did not appear until early in December, mammals around Christmas and man very late on December 31. As a matter of fact, the entire known history of mankind would take place during the last few seconds before midnight.

Conversely, the understanding of certain magnitudes can be facilitated by its translation into time spans. If the owner of an $80,000 American house with a 30-year mortgage of $65,000 at 12 percent annual interest rate were to pay one dollar every hour of every day of his/her life, it would take well over 30 years to pay for the loan.

In his incessant quest for the understanding of time, man has attempted to capture its elusiveness by defining and redefining its significance in a multitude of types and categories. Time has been classified as monochronic and polychronic[1] or objective and subjective, each reflecting to some extent the intimate appreciation and behavioral characteristics of human beings when faced with the inevitability of the passage of time. In a general sense, time has also been defined as vertical or horizontal,[2] vertical time representing tradition or traditional values and horizontal time representing newness and innovation. Thus the analysis of events (whether historical, present or future) can be developed as a function of the interaction of the two types of time factors at any given moment.

On the basis of this classification of vertical and horizontal time, it can be concluded that the explosive physical and technological developments experienced by mankind have heightened the effects of horizontal time, which is steadily becoming more influential than vertical traditionalism. As a result, the ratio of occurrence to time has increased tremendously within the past few decades.

Employing, once more, a time compression technique to illustrate this point, consider that if all history since the birth of Christ were compressed into a single year, the Industrial Revolution would have started around the middle of December.

Because of this explosive sequence of developments, time spans have been affected in a multitude of ways, and society has experienced a broad range of kinesthetic changes. Conversely, man himself has also affected time through his continued adaptation to ever faster life cycle processes, since time is only understood by man in terms of himself and the patterns of change that surround him. Thus man is not only accelerating the pace of his life-style and its evolutionary processes but, in doing so, he is also altering the concept of time.

The understanding of the future is, in effect, much more complex as a function of an occurrence/time ratio than as a mere function of chronological time extensions. Conservatively, from an occurrence/time standpoint, 1980 would be as far away from the year 2000 as 1940 is from 1980. That is why forecasting beyond three or four decades becomes increasingly difficult with each year added to the count, since the time acceleration factor becomes increasingly complex as one moves farther away from the immediate future.

Although the relative time spans affecting shelters and communities develop at a slower pace than those affecting such things as clothing styles and automobile designs, in recent years increased accelerations in life-styles, technoscientific developments and communication patterns have been quietly accelerating the processes of change in human habitats. Presently, a new high rise building is being erected on the site where just over a decade ago the new one-and-one-half-story headquarters for one of the largest architectural firms in the country was built at an original cost of approximately $2,000,000 with properly executed studies of possible growth and change taken into consideration during its planning and initial design. An explosive increase in land value, however, made the original building size/land cost ratio inadequate, making it necessary to build the new structure.

Whether or not these accelerated patterns of change relative to the development of habitats meet the subjective need for human recognition of and identification with the urban landscape is another matter. Traditionally buildings have constituted the landmarks of urban societies. The elimination of these symbols of permanence at ever-increasing rates would profoundly affect the future sense of familiarity, identification and attachment between most inhabitants of urban conglomerates and their surroundings.

This may have more significance than it appears to have at first glance. The clues to temporal values embodied in buildings are essential elements in man's interpretation and understanding of his own past. On the basis of an occurrence/time ratio this goes far beyond the preservation of a few selected historical landmarks. Rapid changes in urban and suburban profiles in North America could be damaging because when the permanence of human shelters is undermined an essential source of sequential values is depleted. Memories (no matter how old or how recent) are fundamental to societies. In certain cultures, all human experience is embodied in a comprehensive concept of present. The Burmese language has no words which signify "yesterday" or "tomorrow," and most events are never truly forgotten, although there is no organized historical sequencing in their remembrance. In practice, their past remains present. And in East Africa, the Chichewa language has two different past tenses: one for those events which continue to influence the present and another one for those that do not.[3]

One of the problems of modern society, however, is the extreme complexity of its dynamic life-style which allows little room for remembrance and contextual comprehension. Modern times are rich in developments and advances but they are also cluttered with facts, new concepts and relationships, all of which must be absorbed by human beings on a daily basis. Modern man lives for an unlimited and accelerated *now*.

As a result, society's use of time is increasingly scheduled and pre-programmed. This is unfortunate, because the scheduled breakdown of time into predefined lapse/task units has an adverse effect on creativity. In effect, time is something which cannot be enlarged or subdivided at will, at least not without due harm to human beings; nor can it be willfully controlled. Many schedules fail because they do not take this into consideration.

Cycles, patterns and other groups of identified past occurrences constitute another often misunderstood correspondence between time and man. The study of the past may be a field open to subjective interpretation, but the vital importance of its analysis lies in the fact that the past actually determines and defines ways of thinking in the present, and these thought patterns, in turn, are employed in shaping the future.

The influence of time on the development of habitats is clearly manifested in the changes of three essential factors: stylistic features, programmatic requirements and mechanical components. Stylistic variations are nearly impossible to forecast, since they are the result of cultural transformations. Also, stylistic developments do not follow simple cause-effect patterns. A material as strong as concrete or steel but as light as styrofoam would certainly revolutionize building design and construction; however, how this material would affect the formal characteristics of buildings would depend on a series of satellite factors that go beyond its physical properties. What if this material could only be used in rectangular slabs and strictly in a horizontal or vertical position? Or what if it possessed significant strength only when exposed to ultraviolet rays?

Most forecasts of stylistic developments consistently turn into mere representations of the "ideals" of existing aesthetic concepts. Programmatic developments, on the other hand, can be forecast much more adequately since they follow patterns resulting from recognized social, environmental, economic and technological issues in more predictable cause-effect ratios and which are much simpler to qualify objectively. The most significant spatial changes in shelters will be those of this latter type. Underwater dwellings, for example, could retain several of the spatial features of modern habitats, but the indoor/outdoor relationship so vehemently stressed in modern architecture would, obviously, have to be drastically changed.

Since the future development of mechanical components to be employed in the construction of shelters will result from the interaction between technological, social and economic variables, it is very unlikely that newly introduced components will succeed unless they are first proven feasible. Novelty alone is not enough to justify the widespread use of any construction product. Buildings are very expensive commodities already. Only the improved spatial adequacy resulting from the reciprocity between mechanical developments and programmatic variations may end up justifying in the future something that is considered unfeasible today. For example, a satellite winter greenhouse-type structure for houses in the northern United States may very well double as a porch in spring and summer or as a portable dwelling to be used for camping. Thus an investment in such a structure, which may appear a luxury by present standards, may prove feasible at a time when the cost of vacation homes skyrocket or workweeks become shorter.

Among the most important mechanical developments in shelters, as far as their impact on living spaces is concerned, will be those dealing with the physical execution, usage and construction of buildings (advances in systems and techniques), rather than with materials or products per se (electronic gadgetry or fiber optics). Although there is a distinct reciprocity between the development of new construction materials and new building systems and techniques, it is the impact of the latter on the former that holds the most significant promise for the future. The "Spanish" architecture of the southwestern United States does not owe its development to the adoption of adobe as building material by the Spaniards, but to the use of that indigenous building material to form masonry units, a construction system ignored by the natives of the region who only used adobe as mass to build their shelters.

Changes in how time affects human beings is also an area which may impact habitats significantly. An increasingly older population base in America, for example, will represent a formidable socioeconomic and political force that will unquestionably reshape the urban and suburban profiles of this nation.

In a generic sense, an older population could affect the social habits and life-styles of this nation drastically. Time is viewed and experienced quite differently by people in different age groups. The fact that time seems to pass more quickly for adults than for children has been related to the capacity for acquiring more or less information within specific periods of time. Mental and leisure activities are directly related to the psychological perception of the passage of time.[4] Everyone knows how quickly time seems to pass when one is psychologically occupied and how slowly it seems to flow when one is bored. An older population (or a combination of more retired persons and shorter work-weeks) may well lead to new developments in fields dedicated to provide for the "free" time of the American population. Occupational patterns, scheduling and entertainment are areas of social interaction that would unquestionably be seriously affected.

Increased leisure time is an area of special interest because it is noteworthy that people's choice of leisure-time activities takes place mostly inside housing units or within a very small radius inside the community. Table 4 illustrates the top six categories of these leisure-time activities and their relative percentages.

There is no reason to believe that housing will ever stop being the basic container of most leisure-time activities. Two of the fastest growing groups of industries in the United States, entertainment and electronics, are aiming their objectives *specifically* at the American housing unit; not at specific households (paradoxically, individuals are a lot easier to classify than household types) but at specific individuals within society. And the only sure way of reaching *all* dwellers is to reach *all* habitats. Personal computers may well follow a similar course. After all, the industry has not labeled them "household computers," but "*personal* computers."

Longer life spans are another generic aspect of the time/age corre-

Table 4
Leisure-Time Activities

Categories	General Percentage Standing	Relative Percentage Standing
Watching TV	41	27.2
Active sports	30	19.8
Reading	24	16.0
Gardening/yard work	20	13.2
Workshop/hobbies	19	12.6
Visiting friends or relatives	17	11.2
TOTAL	151	100.0

SOURCE: Based on data provided by William R. Ewald Jr. in "Part I: Reconnaissance of the Future" of *Creating the Human Environment* (Urbana: University of Illinois Press, 1970), p. 57.

spondence which may reshape the future of American shelters. Presently, for a 40-year-old person, a 30-year homeowner's loan means payments that will last until the retirement age of 65 and beyond (possibly even beyond death, with a 68- to 72-year life span); but if the life span of human beings is increased to 120 years, a 30-year mortgage on a home for a 40-year-old person might be a very common transaction that would end at 70 years of age (less than two thirds of a 120-year lifetime). Changes of this sort would unquestionably affect the location, quality, size and style of edifices throughout this nation.

An older population base would also influence many of the country's sociopolitical characteristics. If retirement ages continue to be pushed back or periods of semi-retirement are introduced, there will be corresponding changes in buildings, communities and urban patterns. Home ownership itself may become a condition based on occupational parameters, with each employment phase having a direct effect on housing, workplace, etc.

The effects of man/time correspondences on human habitats are vividly portrayed in the continual patterns of change of buildings. Change, as a direct result of the passage of time, is an essential part of life. Without change, the passage of time would not actually matter. Moreover, the specific *rate* of change is directly related to the need to understand and plan for the future. The preoccupation with tomorrow is a direct function of this rate, and very significant in periods of vital sociological readjustment such as modern times.

Because of the high level of technoscientific sophistication that has already been achieved, the need to speculate about the future is now more important than ever. In modern times the future is, in effect, now. It is conceived, manufactured, altered and realized now. Man appears to have fully embraced the belief that he can finally master his future. How accurate this controversial assertion is, has been seriously questioned by many. However, it is undeniable that if a better understanding is achieved in the basic areas of human and scientific development that will shape the upcoming years, man will acquire greater control over his own destiny.

The need to understand the future is an extension of the need to understand the present as well. In modern times this task is being attempted by a new branch of scientific research that is aptly called futurism or futurology, and its reports are summarized in what have been commonly labeled future scenarios or forecasts.

11
Futurism and Forecasting

Futurists cannot predict the future any more precisely than anyone else. No one can determine with any degree of accuracy what people are going to do or how they will evolve. What futurists *can* do, however, is describe the logical evolution of trends; that is, they can analyze the probable development of patterns, forecast the likely consequences, and in doing so, provide a *choice*.

Invariably, the crises of today were only minor issues yesterday. Some 20 years ago, scientists discovered that the waters of some lakes had reached acid levels too high to support life. However, well over 15 years passed before water pollution was taken seriously, and as a result 140 lakes in Canada are already dead and officials say that 48,000 more are threatened, 20,000 lakes are dead in Scandinavia, and in the Adirondack Mountains the last count was 212 lakes dead and 256 approaching critical levels of acidity.[1]

Studies of possible developments can prevent future crises because the decisions made today will define the main issues of the decades ahead. In the future, the consequences of present actions can telescope as much as the trajectory of a slight angle variation as it moves away from the vortex. Many world leaders, for example, have complained of their inability to make things happen, and yet many of those same leaders (Harry Truman is one clear example)[2] affected drastically the future of their nations or of the whole world with their decisions.

A very important characteristic of the future is its plasticity. The future does not merely happen; it is *made* to happen. It is not preordained and thus cannot be effectively disclosed; but since it is built upon today it can be rationalized, understood and planned for.

In a sense, everyone forecasts future events. Any plan is merely a course of action based on an expected sequence of events. Forecasts are nothing but an extension of history into the future; they have been defined as ''looking ahead through a rearview mirror.''[3] They are the studies of the evolution of society in a time frame not yet realized. When a developer sets out to build a certain number of housing units in a specific urban sector, he does it as a result of identified needs, market conditions and prospective buyers' attitudes based on *forecast* patterns of urban/suburban growth, financing availability and social evolution. Actually, speculative building and land development rely so heavily on forecasts that even short-term borrowing terms can be significantly influenced by the lender's interpretation of predicted trends, thus facilitating the realization of the forecast from the beginning. The end will invariably be affected by the means. Ultimately, all planned actions are the result of analyses of the future and end up determining the future accordingly.

To establish these analyses of future trends, forecasters turn past data and present conditions into possible future developments utilizing two basic principles of logic: continuity and patterns.[4]

Continuity is that principle which lets anyone forecast the sequence of seasons, aging, day and night, etc. Patterns, on the other hand, refer to recurring formats which do not involve a continuous flow of events: a four-year-old boy may be forecast to grow to six feet, zero inches tall by age 25, but not to twelve feet, zero inches tall by age 50, due to an established human growth pattern.

Among the most prevalent techniques used to develop forecasts are econometric and technological models. Econometric models consist of equations based on economic theories and estimates derived from historical data and relationships, and are employed to forecast the development of existing systems. Technological models are based on engineering probabilities and cost estimates and are used to forecast the development of new systems, the finding of new resources, etc.[5] The results of both models, however, are generally modified by the personal judgment of the forecaster to make the end result more compatible with his overall understanding of the situation. Value judgment factors have been introduced in econometric and technological models in the form of proportions, "accelerators," divisors and multipliers, and in fact can largely alter the end result of the studies so manipulated.

Forecasts can be qualitative or quantitative based on the type of data and analysis report produced, but obviously the end manipulation of the results of any quantitative data as well as the subjective interpretation of qualitative issues produces enormous differences from one analysis report to another.

One pitfall of quantitative reports is their excessive reliance on linear projections. Quantitative data may serve to reveal trends, relationships and reciprocities, but in no way can linear or exponential quantifications serve as the only basis for long-range forecasts. For example, using all the advantages of modern forecasting systems, predictions were made for an increase in housing units in the United States from 69 million in 1970 to 83 million in 1980.[6] However, the discrepancy between forecast and fact (only about 7 percent in actual numbers) reveals that the studies underestimated the actual growth by well over 5 million housing units, enough dwellings to house the entire population of New York City!

Obviously, a classification of all types of futuristic studies could fill several volumes; there are macroanalyses and microanalyses depending on their extent, and optimistic or pessimistic forecasts depending on their overall outlook. However, there is an important characteristic of forecasting formats which should be considered in more detail, and that is the sequential form followed in reporting the occurrence/time ratio. This characteristic may actually divide forecasts into conservative or radical categories.[7]

Conservative forecasts are those in which the future appears to be much like the past with no essential changes in the socioeconomic, political or cultural features per se, but only in their actual rate of modification. This attitude has the disadvantage of underestimating the influence of drastic changes which have traditionally been known to have an enormous impact on the future. Despite very carefully executed forecasts after the world oil crisis of 1973–74, the consequences of the Iranian revolution alone caused the price of oil to rise to levels that 1978 forecasts had not anticipated before the year 2000.

Radical forecasts, on the other hand, project themselves into the future without really developing a path which shows *how to get there*. This type of forecast is dangerous in building construction where transitions are inherently slow due to the basic characteristics of the industry (for example, it is labor intensive and requires large investments) since to ignore the processes of transition to the future is to ignore the basis for the evolution of habitats.

The outline of forecasts varies tremendously from one report to another, and may range from a simple listing of issues to detailed and complex scenarios. Based on a general summary of items analyzed by several futurists during the past decade, a brief forecast of the next 80 years could have the following format:

1980/1990:	Difficult economic era. Population growth pressuring natural resources. Third World nations form cohesive blocks. Shorter work week. Worldwide revolutions. Electronic revolution. Nuclear arms race reaches staggering proportions.
1990/2000:	Average people have portable computers. Artificial and synthetic organs. Underground and earth-sheltered housing proliferates. Weight loss without dieting. Average life expectancy extended to 80 years. First industrial/ manufacturing processes begin in outer space. Space station orbiting the Earth. Explosive growth of robotics.
2000/2020:	Political disorders worsen. Moon bases established. Worldwide political dictatorships. Aquafarms. Worldwide droughts. Mars and Venus are explored. Businesses and government completely computerized. Cure for cancer is found. Factory operations mostly automated. Food is manufactured directly from chemicals. Regulations in bioengineering applications and uses. Average life expectancy 100 years.
2020/2050:	Utopian world government. Colonization of outer space. Computerized management. Life expectancy 150–200 years. Drugs to treat se-

nility and assist memory. Portable/mobile shelters. Vacations on the moon. Worldwide institutions, corporations and currency.

2050/beyond: Vacations in space. Completely automated farms and factories. Robots used for mining, manufacturing and companionship. Superhumans are developed by the use of chemicals, bioengineering, corrective surgery, transplants, etc.[8]

The problem with this outline, however, is that because of its broad layout of issues it does not deal with any particular matter in detail. In fact, one of the most complex areas of forecasting is the translation from generalizations into specifics, the difficult step from macroanalysis to microanalysis.

To develop an adequate link between the analysis of major national trends and their repercussions in shelters and communities, some basic reciprocities between the human shelter and national socioeconomic, political and cultural phenomena must be clarified from the outset. First, it must be pointed out that, economically speaking, edifices represent a durable good of long life and lasting value. Second, most buildings are basically developed to house people as the need arises; they are not the product of a "growth" industry which may feed on itself to develop further, but a response to demands and marketplace conditions beyond the control of any one industry or group of industries. Third, the construction of shelters, whether viewed as an industry or as a technological function, is an area of national endeavor that has consistently shown an overall slower rate of change than many others. Fourth, the development of habitats is extremely susceptible to changes in the national economic situation and has continuously been affected by monetary conditions, occupational fluctuations and regulatory pressures. And fifth, it must be noted that, because of their cost, most buildings must be able to adjust to the variables imposed by the ever-changing environmental conditions, resources availability and social changes consistently experienced by this country. A 1981 F. W. Dodge Report, for example, showed that 77% of the national building activities for the year included repairs, restoration, remodeling, additions or maintenance to *existing* structures.[9]

To organize the analyses relative to the evolution of American built environments, the outline will concentrate only on those specific issues concerning habitats throughout the nation. Thus the forecasts developed will deal exclusively with the areas defined below:

1. *Elements*: Building components; Building types; Functional/programmatic changes; Cities and communities; Urban/suburban rural trends; Transportation/communications.
2. *Systems and Products*: Construction procedures; Building systems; Products and production; Materials and resources; Alternatives and substitutes.
3. *Organizations*: Communities and groupings; National regions; Social and political issues; Life-styles.

Elements are those items that provide a framework for the basic evolution of habitats; *systems and products* deal with the physical execution of buildings and construction; and *organizations* include the socioeconomic and political factors acting upon the development of shelters.

These specific areas of analysis, however, can only be properly defined once the interactive factors affecting American habitats have been identified and their dynamics clearly defined. The basic indicators that will determine all forecasts relative to the evolution of shelters are those which spring from the degree of national wealth and power. It is on those basic factors—economy, resources, technology and society—that the attention must now be focused.

FUTURE FACTORS

Broadly defined, the interactions which will define the future development of habitats in this country can be represented as shown in Figure 3. Ultimately, the reciprocity between any two or three factors will affect the balance of the entire complex.

Each one of these basic factors has special characteristics that may vary according to patterns of growth, scale and territoriality; each one is also influenced by particular elements which give it a relative uniqueness within the complex.

The Economic Factor

Studies dealing with the economic aspects of the building industry can be divided into two distinct types: the general and the particular.

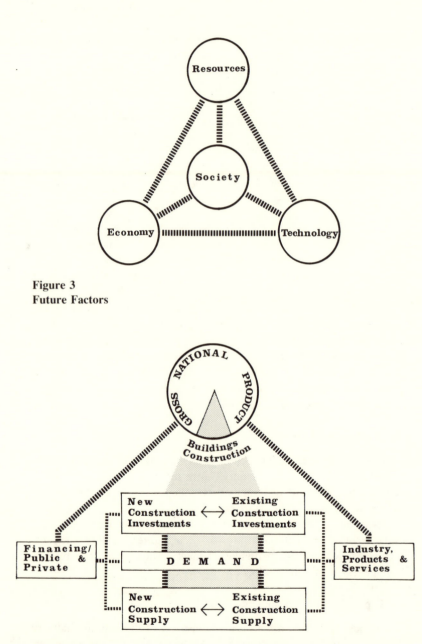

Figure 3
Future Factors

Figure 4
General Economic Factors

The general study deals with national issues (building construction as a proportion of the gross national product, government outlays for housing and construction, and investments by private and financial sectors). The particular study deals with households, businesses and industries, and the financial conditions or real estate transactions related to specific building types and cities.

Both outlooks are extremely useful in laying out a clear picture of the future impact of habitats on the economy and vice versa, since their relative correspondence is continuously affecting the various aspects of the building industry. Graphically, the major interactive components at the general level are shown in Figure 4, while those at the particular level are illustrated in Figure 5.

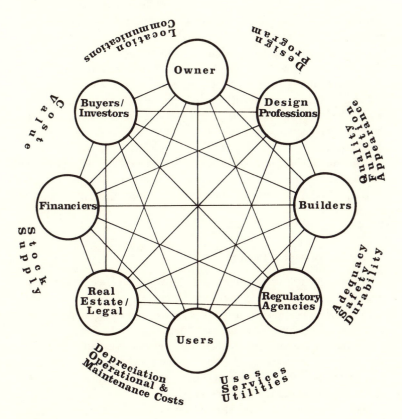

Figure 5
Particular Economic Factors

The Resources Factor

The issues considered here are those dealing with climate, the environment, energy, foodstuffs, land, materials and labor. It must be stressed, however, that the unavailability of certain resources will cause not radical but gradual evolutions in the construction industry. The development of shelters is not an area of human endeavor that lends itself to drastic changes over short periods of time. The worldwide depletion of manganese, for example, may cause a shortage of steel and affect the quality of steel and iron products throughout the nation. However, how that will affect the construction industry is another matter: the use of steel to form structural members in construction is far more important than its plentiful availability for use as building skins. The feasibility of reinforced, prestressed and posttensioned concrete is essentially dependent on the use of reinforcing steel bars or cables. Obviously, a shortage of manganese (with no known substitute to replace it in the fabrication of steel) will have a much more drastic impact on the supply of reinforcing steel bars, cables, concrete anchors and inserts than on the availability of metal panels and steel for automobile bodies, since while there are possible alternates for the latter there are no known substitutes for reinforcing steel in building construction.

Regarding resources availability, the most significant factors will be those affecting the American patterns of water availability and land utilization and development. This is covered in more detail in the next chapter.

The Technological Factor

Here again, it is very unlikely that the world will see an actual decrease in the number of technoscientific developments. In this particular case, the question is whether the present rate of development will accelerate, slow down or remain unchanged.

The basic areas of technological development that will unquestionably influence the future evolution of shelters are: alternative energies, new materials and systems, bioengineering, communications, automation, transportation, and electronics.

One of the most paradoxical problems confronting modern societies is their ignorance of the true processes of technological development.

Traditionally all technology has been evaluated according to three basic properties: possibility, feasibility and safety. However, the most serious problems facing industrialized nations today result from conditions caused precisely by the satisfaction of those same properties. Aluminum cans have been proven *possible, feasible* and *safe* but they constitute a serious problem in parklands throughout the nation.

Scientific developments are also a complex area of study because an excessive confidence in positive evolutions is dangerous in that it may cause remedial movements to stagnate, while overly-concerned negative views of science and machinism may cause damaging extreme reactions. One observation made previously should be reiterated here: technological revolutions *are not* the product of identified social or economic needs, but they *will* change the behavioral traits and evolutionary patterns of societies.

The Sociological Factor

The major areas of concern here are world affairs and their impact on domestic issues, government and population growth, crime, styles of living, entertainment/leisure habits, and the evolution of several sociological elements such as family and age groupings.

While sociological changes are by far the most complex of all the factors governing the future, they will rarely involve only one or two isolated issues, but, like resource availability, will tend to form complex webs which will act upon the development of habitats only gradually. Ultimately, the effects of many technological advances relative to the development of shelters initially follow a pattern of passive integration and do not affect spatial layouts and construction significantly until their full sociological implications have taken hold.

The most outstanding discoveries and applications of modern science, for example, have taken place in buildings erected several decades ago. University buildings are among the edifices most likely to maintain traditional facades while remodeling their interiors to fit spatial requirements of the most advanced nature. Not long ago, the program for the basement remodeling of the natural sciences building of a leading university required the creation of one experimental laboratory with six chambers in which environmental conditions ranging from desert-like climate to fog, rain or even snow/frost could be reproduced. Such requirements would be complex enough to meet in any type of structure, but in this case the difficulties were aggravated by two other factors: the building was fairly old—originally built in 1910—and the basement had only a 7.5-foot clearance from its concrete floor to the underside of the structure above it. The extreme complexities of that particular project are far too numerous to mention here; however, one point is worth mentioning: the project *was* successfully completed.

The spatial changes and functional readjustments forced upon habitats by modern sociological changes have unquestionably altered the original design intent of many edifices. This is especially true of large institutional buildings and of a significant number of commercial and public complexes developed around the turn of the century. However, in spite of this apparent loss of spatial integrity, when looked upon contextually the changes introduced by progress have consistently brought American habitats closer to environments that are essentially defined in terms of man's functioning. Despite the terrible conditions that plagued most American and European cities during the Industrial Revolution, their overall environments represented an improvement over the conditions prevailing in most urban nuclei throughout the preceeding decades.

In the same light of this retrospective point of view, it would not be surprising to find contemporary houses evaluated in the future as "exceedingly large dwelling units cluttered with useless items, ineffective appliances and rudimentary materials, all distributed around oversized rooms that required about 50 to 60 hours of physical labor per week to operate and maintain and another 120 hours of work per month to acquire and support."

12
Economy, Resources, and Technology

WORLDWIDE ISSUES

Increased political instability and unchecked population growth are the two most significant factors to be considered in the future development of worldwide issues, since they will have major impacts on trade patterns, resources availability, migratory trends and economies throughout the world.

If worldwide political stability could be achieved within the next two or three decades, it is possible that international cooperation in technological developments, resources distribution and population control could further strengthen worldwide economies. However, this could affect this country negatively, since with only 10 percent of the world population the United States consumes 45 percent of the world's resources, and any equitable distribution of commodities would require a decrease in the lion's share presently held by this nation. Hopefully, technological developments could offset these conditions somewhat.

Regarding the development of shelters, international cooperation could improve the housing, building standards and urban problems of under-developed nations significantly by exporting technologies presently available in the industrialized world, since the housing crises and urban growth patterns presently exhibited by Third World nations are very similar to those experienced by this nation during the years of the Industrial Revolution.

Worldwide population growth adds 200,000 human beings per day to the population base.[1] Urbanization in the Third World is a serious matter. Every day 75,000 people migrate from rural areas to cities. In Mexico City alone the growth has been—and continues to be—explosive: from 1 million inhabitants in 1930 to 14 million in 1980 and a forecasted 32 million by the year 2000.[2] Not surprisingly, this oil rich nation is undergoing one of the most devastating economic crises in history.

Excessive foreign population growth has an indirect impact on American habitats since crime, unemployment, overcrowding, riots and other maladies lead many foreigners to abandon their exploding urban areas and relocate elsewhere.

Despite the positive effects that worldwide cooperative technoscientific developments would have on economic and sociopolitical welfare around the world, however, efforts in this direction are highly unlikely to occur for at least two or three decades. Thus the worldwide economy is most likely to go through a relative decline that will affect balance of payments and trade of commodities during the next 20 to 25 years and experience a gradual tapering off towards the beginning of the 21st century, once the full impact of the global population growth on world resources has stabilized.

Because of this, American industries and businesses dependent on imports from the Third World may find themselves at the mercy of the ups and downs of unstable worldwide economic and political climates. Moreover, the substitution of certain raw materials may become essential, if not because of economic instability, simply because of their limited availability. This could affect the evolution of shelters in the United States drastically. American industries employing nonrenewable resources that are directly related to building construction consume over one-third of the entire world's production of aluminum and nickel and one-quarter of the globe's chromium, copper and tin production.[3]

The consideration of most characteristics and trends of the worldwide socioeconomic and political situation determines three key factors directly related to the future development of American habitats: first, during the next decade the United States will most likely expe-

rience increased difficulties in obtaining specific raw materials and resources from some overseas markets; second, this nation can expect
larger numbers of foreign immigrants as more people flee the declining
environments, unstable economies or the political upheavals of impoverished nations; and third, a resulting increase of foreign investments
in this country as most overseas wealth tries to find economic and political stability.

ECONOMY

American economic climates are, in effect, reflections of a multitude of interactive conditions far too complex to establish in detail here.
However, in considering the relative ups and downs of the American
economy as a whole and the way these cycles relate to American habitats, parallels can be drawn between socioeconomic conditions and
building construction.

In a flourishing economy, habitats would improve in quality and durability; the nation would witness a growth of small investors and developers (a trend which would affect regional land use patterns); and
houses and buildings would maintain a high square foot per person ratio.

Also, a flourishing economy would mean that households would
continue to increase their use of appliances, electronic equipment and
entertainment gadgetry. But a flourishing economy would also prevent
this nation from achieving effective or long-term energy conservation
objectives, waste control and historic preservation goals, as well as increase many emerging environmental problems (deforestation, pollution, land pillage).

Other significant consequences of a flourishing economy would also
be increased population mobility; continued migratory changes, to the
sunbelt regions; higher housing and building costs; increased suburban
sprawls; higher real estate valuations and property taxes; and increasingly higher costs of utilities and other supporting services.

Beyond the early 21st century the practical effects of a flourishing
economy on American habitats in general are difficult to outline; however, some possibilities can certainly be evaluated: there is a physical
limit to suburban sprawls and that limit will be defined by public utilities and land cost factors. It will be increasingly difficult for this na-

tion to maintain its present rapid suburban expansion patterns at the expense of arable lands and forests. Thus the growth must develop along different courses.

Moreover, many of the urbanization problems that the United States will most likely face in the coming decades will occur in precisely those areas that appear to be most desirable today: namely, those in the sunbelt region. Paradoxically, at the dawn of the new century the sectors of this country that will have conquered most urban decay problems will be precisely those that seem so unappealing today because of their urban maladies: the Northeast, the upper Midwest and certain south-central regions. In addition one must point out that no matter what technological advances may occur through the early 21st century, the relative location of those parts of the country with respect to Europe, Africa, the Middle East, Scandinavia and the Canadian borders will remain the most strategic.

Conversely, a declining economy in the years ahead would produce a reduction in the quality of building materials and construction and result in increased operational and maintenance costs. It would also accentuate the difference between the levels of quality among buildings and thus cause jolts in urban/suburban development.

The high cost of money is one issue which will drastically affect the future of most American cities and towns. With high interest rates, the housing market *will* crash unless home ownership is transferred from the individual to the corporation. By present standards, any figure over a 12 percent interest rate will absorb well over 60 percent of the net earnings of any individual for the cost of shelter alone. The average monthly budget for first-time home buyers in the United States in 1981, for example, expressed in terms of percentage of household income, was figured to be as shown in Table 5.

So with the combined total of rent and utilities already consuming nearly half of the household income, any further increase in financial costs would probably eliminate home ownership for much of the population. At 1981 inflation rates, an average home of approximately $60,000 would cost about $1,643,000 in the year 2000.[4]

A declining economy would reduce the percentage share of detached single family dwellings in the overall housing spectrum. Moreover, a depressed economy would probably affect the usage pattern of many existing detached single family dwellings, causing changes to accommodate private gardening and in-house agriculture; more dwell-

Table 5
Average Monthly Budget Distributions

Item	Percent
Rent	35
Food	22
Utilities	9
Entertainment	11
Savings, Payments, Credits & Investments	23

SOURCE: Based on data provided by Richard McNeilly (ed.), "Who They Are," *Professional Builder* (January 1981): 205.

ers per unit in increasingly complex household structures or boarding/rental practices; housing and building co-ops; fewer privately owned automobiles, appliances and amenities (garages, storage areas, swimming pools, etc.) per dwelling unit; and the return to inner city areas for many suburbanites. A decaying economy would also delay building improvements and urban renewal projects, which would in turn cause higher population concentrations in older urban sectors and in small cities which may not be ready to handle the increased densities.

These economic declines could cause the gradual disappearance of the small builder and the small development company and give birth to a new type of organization: the mega-builder, a corporate structure capable of providing the nation with a comprehensive approach to the supply of shelters (design, construction, financing, maintenance, security, insurance and real estate services). However, this trend, which may eventually provide the solution to many problems posed by the inadequate supply of housing, might cause the further standardization of building types and an increase in corporate owned, supplied and controlled habitation.

Obviously, a declining economy would cause a relative reduction in population mobility, although due to different regional characteristics this might not be significant unless the economic decline became so depressed that it forced occupational "status quos" throughout the nation.

Since it is highly unlikely that the national economic panorama will change drastically, one can count on the continuation of cyclical periods of economic improvement and decline through the next millennium. The periods of economic decline would accentuate energy savings and operational efficiency efforts, facilitate the further development of entertainment and leisure type activities at home; and produce smaller shelters (or larger houses destined to accommodate bigger households, but always with a smaller square foot per occupant ratio) that will become more utilitarian in nature and allow for multiple uses for traditionally programmed areas (such as dining area, study, workroom and living/family room). Economic decline might also force an upswing of the shared ownership and maintenance concept for certain building amenities such as parking areas, gardens and lawns, pools and tennis courts, and might strengthen the concept of joint/cooperative home ownership arrangements by specific household types: retired couples, single parents, homosexuals, tradesmen, etc.

The periods of economic decline would also increase the number of do-it-yourself features in habitats, and this is one dangerous trend that is likely to develop further in the coming decades. There are two serious cautions to be made here: (1) real estate valuations become unrealistic or misleading when certain features of a building have been executed by amateurs who lack the tradesmanship quality of a professional builder, and (2) the dangers posed by structures built by inexperienced hands and their lack of certification could be a matter of serious impact in future habitats and may end up creating a whole new wave of regulatory constraints that could further damage the evolution of American living units.

Regarding the spatial program of most single family dwellings, it is noteworthy that more activities of the do-it-yourself type would increase home workshop type rooms, introduce new storage requirements, and cause new safety concerns and insurance premium increases.

A declining economy would most likely result in an increase of ''prototyped'' buildings, serialized housing, and the gradual decay of shelters and buildings in certain urban and suburban sectors. It is also very likely that in this scenario mortgage arrangements would become more complex, and be further altered by factors such as increased life expectancy and social mobility. Because of a slumping economy during the early 80's, several innovative mortgage types were introduced in the real estate home ownership fields; such as shared appreciation contracts, variable interest rate mortgages, land contracts and deferred payment stipulations. Few of these new types of mortgages, however, have become widespread enough to undermine the fixed interest rate long-term mortgage as the most common system of home financing throughout this country. In this respect, the latest innovations appear to be more in new kinds of firms entering the long-term mortgage field than in new types of mortgages.

On a national scale, the major economic factors affecting the evolution of habitats can be defined in terms of six basic conditions. The first is the basic correspondence between the development of shelters and energy, maintenance and operational costs. There will unquestionably be a close relationship between the American habitat and its energy/operational factors.

Second, the American shelter will continue to experience a boom in entertainment/leisure type activities due to further developments in automation and the modernization of life-styles and occupations, higher costs of away-from-home entertainment facilities, explosive growth of the industries that provide in-house recreation, and broader communication patterns.

Third is the basic need to reduce the relative costs of shelters as a percentage of personal income in order to maintain adequate housing either through new technological and scientific advances or through spatial curtailments, certain work-at-home patterns, etc.

Fourth is that there is a limit beyond which suburban sprawl can no longer be considered a viable answer to most urbanization/ suburbanization problems. Although that limit has not been reached yet, it is very unlikely that unchecked suburban sprawl can continue far into the next millennium.

Fifth is directly related to the traditional mobility that characterizes American life-styles and particularly the American work force. The population of this country will probably continue to be increasingly mobile and this will have a serious impact on its habitats. Somehow the inherent discrepancy between shelter permanence and social mobility will *have* to be resolved. This in fact, could be one of the most essential requirements of future shelters.

And sixth is the future growth of the sunbelt regions in the next few decades. These are the regions most likely to absorb much of the nation's population increase in the years ahead, but they will eventually be confronted with many of the same types of problems that presently plague other highly urbanized regions. Those regions, in turn, may well experience new periods of boom and prosperity due to their decades-long population stability, their resolution of urban/suburban problems, and their favorable locations.

RESOURCES

Improved environmental conditions would help preserve the integrity of most materials and exterior skins in American habitats and provide better built environments for the American people. Environmental deterioration, on the other hand, could accelerate the consolidation of different functions/activities under one roof (working, sleeping, eating, recreation) to provide day or week round indoor occupancy and increase urban decentralization and suburban sprawl. Paradoxically, many environmental issues end up conflicting with one another.

Since detached single family homes are the least efficient type of housing unit, further deterioration of the American environment in the form of acid rain or climatic changes would also affect their maintenance and operational costs more severely than town homes and multifamily buildings or commercial structures.

The distribution of some basic resources could be another area of concern. Water shortage, for example, is destined to become a very serious issue in many parts of this country as early as the late 80's. It has been forecast that several key aquifers—such as the Ogallala Aquifer which extends from Nebraska to Texas—will run dry in 40 years.[5] And in California, a recent referendum voted down a proposed water supply project for the southern part of the state: the Peripheral Canal. This project, originally estimated at 7 billion dollars and delayed by red tape and environmental issues for about 10 years, ended up tripling its original cost to 21 billion dollars and was eventually defeated at the polls due to its high price tag, the voters having decided in favor of several water conservation measures of questionable adequacy. Considering that even with all the advantages of modern technology and building/highway construction, a project of the magnitude of the Peripheral Canal would take a minimum of 20 to 25 years to develop, it is apparent that for some areas of the country it may already be too late.

Problems involving resource distribution would affect not only the cost of utilities (water, sewerage, etc.) but the production of many foodstuffs in the affected sectors. Such a decline in the production of foodstuffs could force the American shelter to incorporate private year round vegetable gardening and possibly even some small animal farming in certain suburban sectors. Average home gardens, for example, are 595 square feet, cost about $19.00 and yield a value of approximately $325.00.[6] It has been argued that eventually, urbanized sectors could end up producing more foodstuffs than farmlands through advances in climate control and food-producing techniques, but such theories look too far into the future to be considered likely occurrences in the forthcoming decades.

Presently numerous technologies are being developed that may help remedy negative food supply conditions, such as new methods for preserving foods without refrigeration, hydroponics, synthetic food production and processing, and an array of systems and techniques that permit small animal breeding in captivity (i.e., rabbits and shrimp). However, the changes that could result from the incorporation of some of those technologies into private dwellings would have complex and highly dangerous ramifications involving occupational and health hazards, regulatory controls and enforcement.

The relative correspondence between the development of shelters and the environment is another area that will merit close watch in the coming years. Except for controlled suburban sprawl and changes in the water supply and sewerage/waste disposal of habitats throughout this country, there are few areas in which innovations in building technologies and operation can help prevent environmental decay significantly.

In response to unfavorable climatic and environmental conditions, American habitats have developed minor defense mechanisms such as special purpose coatings, artificial climate improvements, new impervious materials, concrete admixtures, and others. The most significant changes in the American built environment of the last decade, however, have been caused by the awareness of a new essential requirement in the development of modern habitats: energy efficiency.

Energy

Declining energy costs would be unlikely during the next few decades since finding substitutes for the present energy sources is a long and complex task. In all likelihood, the cost of energy will remain high and thus have a significant impact on shelters.

Higher energy costs could multiply the regulatory measures concerning energy usage, and could result in intensified energy conservation efforts in the design and construction of shelters, passive and active solar technologies and the retrofitting of older buildings.

While the search for energy substitutes continues, certain national regions will most likely flourish more than others. Obviously Alaska, Texas, Montana and Oklahoma are bound to experience higher economic growth than other states, but as improvements in solar technologies, nuclear fission and other alternative energy sources are developed and more and better energy efficiency measures are incorporated in the design, construction and operation of habitats, regional economic balances will probably be reestablished.

Current trends appear to be aimed at turning the single or multifamily dwellings of this country into self-sufficient units that produce and regulate their own power sources, provide most of their utility needs, recycle their wastes, and operate pretty much like "automated" shelters. However, although most of the technologies necessary to achieve the majority of these goals are available today, their implementation on a national (or even regional) scale is another matter, since the capital outlays necessary to incorporate these advances are positively staggering in retrofit costs alone, not to mention the resulting decline in costs of abandoned utilities (which could, in turn, limit the feasibility of new retrofits), the legal ramifications, or the enormous losses of public and private investments in traditional utility services and related industries.

A strong possibility, however, is the *gradual* abandonment of existing utility systems. As we will see later on, electric toilets and efficient recirculating-purifying water systems may reduce the water consumption and sewerage disposal of housing complexes to such an extent that only small supply lines would have to be provided to furnish water and no sewer connection would be necessary at all. This would help ease present strains on utilities significantly since the sewage produced in the United States stands presently at 21 billion gallons per day (about 100 gallons per person).[7] By the same token, all electricity, heating and cooling could be generated by each dwelling individually through advanced photovoltaic or insulation techniques. These kinds of developments would mean that as newly developed areas incorporate such advances, edifices located in older sectors, where public utilities are still available, would have the option of employing traditional systems or completely renovating their utility systems (from public to private) by incorporating some or all of the new technologies available.

Land

This is one area that will pose serious problems unless proper measures are taken to control its diminishing adequacy and availability. Habitats throughout the world may be seriously affected by what will probably be a worldwide land crisis sometime in the early 21st century. In 1950, approximately 25 percent of the Earth's land surface was forested; in 30 years that figure declined to less than 20 percent. Every year the world loses areas of forested land equal to the entire area of the state of Indiana. The situation may appear to be less critical in North America, which has the third largest total of closed forest areas in the world (approximately 25 percent of the total land area); but this average figure includes Canada, a country with a small population base and extensive forested acreage. The United States specifically has lost one out of two acres of its woodland since 1950 and still loses about 35 million acres per year (an area about the size of Cuba).[8] In this country the total land base of 2.3 billion acres was mostly dedicated in 1969 to agricultural use, with 525 million acres dedicated to non-agricultural uses. Of these, 61 million acres were urban settlements, 81 million acres recreational and parkland, 27 million public installations and facilities, and another 28 million were miscellaneous areas such as deserts, tundras, and marshes. However, of the total amount of agricultural acreage reserved by the U.S. government in 1969 (140 million acres), only 35 million remained available in 1977 since, of the 76 million listed by the Department of Agriculture as "probable" agricultural land, over one-third was already marked as subject to possible urbanization.[9] This was a likely occurrence, since the United States urbanizes about 2 million acres of land per year.[10]

Between 1967 and 1975 alone, the American forest acreage was reduced from 445 to 375 million acres[11]—a total reduction of 70 million acres (an area approximately equal to the entire state of Arizona) in less than ten years. The problems involved in land availability are far more complex than mere urban sprawl and public utilities expansion. Deforestation and tracts of land lost to mining and other industrial uses also account for significant losses of the American land base.

Land availability in the United States is a serious problem when its impact is measured in sectors such as agriculture, mining and forests. However, from a strictly urban and suburban point of view, land availability is not a critical issue. As we have previously seen, the United States has one of the lowest ratios of inhabitants per square mile in the world. Obviously land, as a habitation resource, is a relatively plentiful item in this country; the problem is that the land presently used to accommodate suburban sprawl is the *wrong kind* of land. If urban growth patterns persist, then the American land base for the production of foodstuffs may be strained to a point where it may, in turn, have a serious impact on the costs of land for urban/suburban developments and deforestation practices.

This is an area of complex economic considerations, since the feasibility of urbanization of farm lands depends not only on the demands of urban expansionism but also on national agricultural policies which continuously affect the value of crops and farms, the national economic stability and, in the end, the intrinsic value of the land itself.

Further decreases of land availability would force vertical urban growth (above and underground), smaller lot sizes and, consequently, an intensification of multifamily type dwellings. The lack of new land developments would also cause real estate valuations and taxes in most urbanized areas to rise and restrain the spread of new communication arteries, thus increasing the traffic problems and congestions of swollen urban/suburban areas.

This could cause many new underground habitats to become extremely popular, since they would allow for the full use of green spaces and in many ways respect rather than modify landscaping and terrains. This trend would also be reinforced if the need for private vegetable gardens and small animal farming became widespread. Underground habitats, however, have some unique problems that—as we will see in future chapters—may prevent their development from becoming common practice, even in extreme cases like those previously described.

Objectively evaluated, the problems of shrinking land bases, soil economies and deforestation practices prove that there is a need to achieve as soon as possible a positive interaction among all the elements that are eroding the American soil. It is doubtful, however, that this may take place before the turn of the century.

Materials and Labor

A continued abundance of construction materials would help stabilize housing costs and improve the quality of habitats by the introduction of new construction systems and products. This development, however, is not very likely.

All present signs seem to indicate that many future construction technologies and products will be aimed at recycling or developing substitutes for traditional building materials and procedures.

It is common to speculate about material shortages triggering the development of new and revolutionary products and systems that will change edifices overnight. However, in an industry of slow evolutionary processes such as building construction, this attitude begins and ends in unfounded speculation. The probable shortages of certain construction materials, however, are not mere speculation, but issues that could affect the construction industry and several building types, specifically, in a short period of time.

Since these shortages would be reflected mainly in higher costs, there would probably be continued increases in the prices of steel, special alloys and coatings, electrical wiring, plumbing fittings and glazing. However, since the use of plastics will, in all likelihood, continue to increase, it may alleviate these conditions somewhat.

A decline in the availability of certain materials could also cause a rise in many real estate valuations and taxes and an increase in the number of preservation and restoration projects. It could also produce new regulatory restraints as new systems and product substitutes were introduced in the construction of American habitats. In a sense, shortages of some traditional construction materials could have the advantage of forcing the construction industry to take a hard look at the need to improve its methodologies and research, and lead to the incorporation of new products and systems in the building process.

One factor acting against construction progress, however, would be the increased availability of manual labor for most building trades. Although available labor could contribute towards reducing building costs somewhat, it will most likely cause many construction systems to stagnate by facilitating the continuation of the present labor-intensive methodologies used in construction throughout the country. One of the key factors in reducing the cost of shelter construction lies precisely in reducing its intense labor dependency rather than in merely cutting labor costs.

Increased foreign immigration would be a factor that could increase labor availability during the next two decades. However, it is highly unlikely that labor supply alone will become a controlling factor in stabilizing the projected higher costs of construction through the next 20 years. The basic problems also include the availability of materials, financing, utilities and, above all, the market demand pressures which may be exerted on the inherently slow process of building construction. It is to be hoped that these pressures will promote the mechanization, prefabrication and "soft automation" of building systems and industries through the next few decades.

This brief analysis of resources, therefore, produces a set of five essential determinants for the future evolution of American habitats. The first is that there is an urgent need for the development of improved food production systems to help deal with the projected worldwide population growth of the next decades. Many of these new systems could, consequently, cause changes in habitats through the possible implementation of new technologies in private dwellings or through the creation of new building types destined to facilitate the production of foodstuffs with artificially controlled climatic and environmental conditions.

Second, there is a definite trend to make the human sheltered environment a self-sufficient entity in regard to its energy production and management as well as its service and maintenance needs.

Third, problems involving land development, deforestation, energy issues and urbanization patterns may lead to the exploration of new frontiers (practically ignored until now) for future American habitats (oceanic developments, underground shelters, etc.) and to a general transformation of the traditional concepts of "shelter." As a case in point, it is noteworthy that the greatest sources of manganese and cobalt presently known are all under water, so to overcome a shortage of these materials, underwater exploration and mining would have to be fully developed.

The fourth point is that a probable decrease in the availability of many raw materials during the next two decades may drive up the cost of many traditional building systems, construction products and edifices in general. But this trend may also lead to the development of product substitutes, improvements in inefficient or wasteful technologies and the discovery and exploitation of new sources. For example, as early as 1977 a study conducted at the University of Hawaii concluded that the outer space mining of a metallic asteroid having a volume of only one cubic kilometer could supply the world's demands of copper for 10 years, iron for 15 years, nickel for 1,250 years and cobalt for 3,000 years.[12]

Finally, the probable increase in labor availability, aside from acting as a brake on the full implementation of advanced technologies, could affect the quality of habitats throughout the United States significantly, since expansions in the population base without adequate supplies of materials or the quality control made possible by the implementation of new advances in industrial production would cause an upswing of unqualified labor forces entering the construction industry.

TECHNOLOGY

One of the most significant factors determining the impact of tech-noscientific developments upon this country's habitats will be whether technology concerns itself with the resolution of problems or continues to follow its traditional pattern of developing the unnecessary. The spatial distribution of a typical 3 bedroom, 2 bath house of approximately 2,000 square feet, for example, allocates 40 percent of its area to the technology intensive characteristics of American life-styles. The cost of these spatial assignments as reflected in construction dollars is enormous, but that is only one of many expenditures that also include utilities, service and maintenance, and insurance costs related to these areas. Table 6 shows the average 1981 yearly cost to operate common household

Table 6
Average Operational Costs of
Household Appliances

Appliance	Cost
Water Heater	$215.17
Freezer	92.82
Clothes Dryer	50.64
Air Conditioning	47.30
Range/Oven	37.23
Dishwasher	18.51
Garbage Disposal	0.36
TOTAL	$462.03

SOURCE: Based on data provided by Richard Smith, et al. *The Average Book* (New York: Rutledge Press, 1981), p. 55.

appliances. Thus before speaking of golden technological futures one must really pause to consider the words very carefully.

There are positive signs, however, which seem to point to some basic improvements and collaborative efforts in the area of social evolution and scientific development, as seen in newly established forums of physics, medicine, engineering, electronics and education. Many of the traditional gaps between social attitudes and scientific communities seem to be disappearing due to gradual improvements in communication that allow scientific advances to keep up with other developments as they occur, and that provide society with an opportunity to react to most scientific discoveries prior to their implementation. Thus man may be beginning to exercise control over his own technological future for the first time in history.

It is likely that modern communication patterns will develop even further, and this would affect habitats on every level: nationally, improved communications would facilitate the mobility of work forces throughout the country, while on an urban scale, they would benefit all forms of transportation by reducing physical travel. This could also lead to improved land utilization patterns.

Individual dwellings could undergo significant changes in a few decades due to new work patterns, changes in the educational framework, or even different socializing practices. Some of these trends, however, also give rise to concerns of increased social isolationism, loss of privacy, lack of individualism and diminished physical contact among people. Much has been written about the danger of widespread electronic communications involving such things as two-way audio-visual/computer hook-ups that would reveal finances, political opinions and banking records; monitored work stations that could reveal employees' performance, abilities and disabilities with the precision of fractions of a second; and increased regulatory measures arising precisely because of these situations. Clearly there are still many questions that remain unanswered.

Among the most significant technoscientific developments for the improvement of American shelters would be those involving building materials and systems. New construction products could reduce maintenance requirements and increase the durability of modern habitats, while improved construction techniques could easily reduce the labor dependency of building construction, reduce building costs and provide better quality control.

Some improved construction products and systems, however, could cause problems in future built environments. The popularization of do-it-yourself type shelters may result in hazardous building conditions or in increased regulatory constraints. New products could also accentuate the already excessive dependency of structures on industrialized products, and end up creating new environmental problems as untested products enter the construction marketplace.

Research and development for the building industries is vitally needed. If this country continues to build edifices with present technologies and materials at the rate necessary to accommodate its inevitable future growth, it will unquestionably face crises having the most significant consequences throughout its urban nuclei, suburban developments and even rural communities.

The result could be a scenario in which the reduced availability of specific construction materials, increased dependencies on foreign imports, and greater deforestation and strip mining efforts could seriously undermine the American economy and environment. It would also result in more complex urban and suburban developments and in a marked tendency towards standardizing materials usage and prototyping edifices and building types. Moreover, the effort to provide more shelters with dwindling resources in a rapidly changing society could set back the exploration of other environments capable of development for human habitation, such as underwater, underground or outer space communities.

There is also the possibility of other industries taking over the development of shelters and eventually of the entire construction market. Mobile homes and motorized shelters, for example, are not manufactured by ''traditional'' construction companies. If the American electronic, petroleum/energy and aerospace industries continue to expand the scope of their operations, future American shelters could come precisely from these sources. This theory is also reinforced by the fact that traditional land-use patterns cannot possibly continue indefinitely in America, so that other industries may be forced to involve themselves with the development of *un*traditional habitats in *un*traditional living environments.

Accordingly, the basic factors that surface from this brief analysis of the possible technological developments concerning the evolution of American habitats are as follows: First, shelters are consistently being modified by the introduction of new technologies within their envelopes. Presently, however, due to the large number of edifices already built, such incorporation/retrofit presents a rather unique problem which is further complicated by the inherent high cost of building construction.

The second factor is the significance of the electronic equipment to be introduced into the living environments of this nation. In housing, for example, the pattern followed could be very similar to that of other state-of-the-art appliances: new developments offering "optional" electronic features with the "optional" feature eventually becoming standard equipment.

Third, there is a marked trend towards achieving self-sufficiency in most American shelters.

Fourth, there is vast room for improvement in American habitats and a pressing need for intensive research and development programs involving building products and systems. The future design and construction of shelters in this country is, in fact, an open field and *will* become the domain of whatever industry or group of industries develops the most adequate living environments for this nation's future.

And finally, today America is in an excellent position to regulate its own technoscientific developments. What this nation will do with its technological future, however, depends not only on economic conditions, resources availability and scientific roles, but primarily on social factors.

The future development of these social factors is the subject of the following chapter.

13
Sociological Issues

The evolution of American habitats is affected by a multitude of social changes. It is in the transformation of life-styles, cultural features and political issues that the characteristics of edifices are truly defined.

It is also in precisely those areas that this nation is undergoing some of the most significant changes it has ever experienced.

LIFE-STYLES

One essential factor that is likely to affect habitats in the coming decades is the increase in leisure time. An important consequence of this on a national scale would be the accelerated growth of regions that offer adequate environments for entertainment and leisure type activities. Although the ripple effects of these developments could reach every corner of American habitats, shorter workweeks, for example, would surely mean improved urban traffic patterns (especially if they are combined with staggered work schedules), adequate mass transit systems, better communication arteries, and improved utilities, services and energy management. But it is noteworthy that if present urban decentralization trends persist, increased leisure time could force drastic changes in many urban commercial establishments, since a shorter workweek would cause most to lose nearly 20 percent of their public exposure.

The continuation of present urban decentralization trends added to a reduction in workweeks would also mean complex communal/suburban activities, functions and services as well as further urban decay.

A reduction of urban decentralization patterns, on the other hand, could help resolve many problems posed by growing leisure-time activities and actually help achieve the rebirth of many North American urban nuclei.

An increase in leisure time would probably cause a boom in the development of vacation housing and resort areas. In fact, the tendency to rent small dwelling units near work centers during the workweek and own a "long-weekend" dwelling elsewhere could well become a widespread practice for most of the American work force in the early decades of the next millennium. Furthermore, increased leisure time could also cause a proliferation of new entertainment-oriented building types and mega-complexes, all destined to help the American population occupy their expanded leisure time (it is noteworthy that most of the complexes of this nature presently operating are already thriving businesses).

With regard to individual dwellings, there could be a marked increase in in-house type entertainment activities as a result of shorter workweeks, and a probable boom in home "entertainment centers" facilitated by the widespread use of home computers, videotapes and other electronic devices. There could also be increases in home workshop, hobby, maintenance or repair activities and a proliferation of many home originated and operated business activities as part-time or even full-time occupations. These trends could not only significantly alter the spatial requirements of most future American edifices, but could also cause drastic changes in zoning restrictions, land use patterns and neighborhood characteristics, since even the partial return of home business type dwellings (an electronically modified version of the pre-Industrial Revolution rural cottages) would radically change the placement of many urban and suburban dwellers and activities.

If this nation were to change somewhat the pattern established during the Industrial Revolution, when workers left their homes and went to work in factories, in order to return to certain "in-house" occupational activities, the effects on urban and suburban America would unquestionably be dramatic. Such movement may not be too far into the future. A recent study sponsored by the National Academy of Sciences, for example, has concluded that by the turn of the century, 40 percent of American homes will have videotex units (a receiving/transmitting unit that combines video screen, printed word or text and computer controls) which will usually allow two-way audiovisual communication channels to become widespread and thus greatly facilitate varied work-at-home activities.[1]

Just as the Industrial Revolution clustered workers around assembly line complexes, the electronic revolution could decentralize many functions, operations and individuals and thus redefine some of the traditional collective patterns of association followed by this nation to date. A back-to-the-home movement, based on a network of steadily improving communication patterns, is likely to be one of the most significant factors that will reshape several elements and occupational patterns of the American society in the coming years. Moreover, this movement could also reinforce one issue noted previously: the trend towards American shelter automation and self-sufficiency, with emphasis on the do-it-yourself maintenance and servicing capabilities of the dwellers.

Increased work-at-home patterns, however, do not fit all occupations nor have they been fully endorsed by most forecasts. Arguments against work-at-home activities maintain that in general people want to be with other people in occupational settings and that physical interaction, socialization and feelings of imprisonment or isolationism respectively could work against widespread work-at-home practices. In the end, work-at-home activities may be limited to certain occupations or only available in some very specific cases.

Conversely, many of these developments could also end up affecting the patterns followed by the American work force. Increased occupational fragmentation and specialization could trigger the development of complex social groupings. Occupational fragmentation could produce a highly qualified work force, but it could also increase the number of special interest groups, cause intricate socioeconomic patterns, and introduce new specialized building types and functional groupings, thus forcing further standardization of spatial layouts in general.

It is noteworthy that any further fragmentation of the components involved in the construction of edifices would hurt the American building industry, since many of its basic problems are caused precisely by the extensive breakdown of its fundamental activities. Moreover, the readjustments in traditional structures necessary to accommodate new specialized functions would contribute to the acceleration of active construction cycles, which are inherently very costly processes.

These issues may be further aggravated by the functional adequacy of most buildings—especially if some single family dwellings develop into full- or part-time work centers—that could be in direct conflict with the mobility needs of drastic occupational changes. Also, since it is highly unlikely that many people would remain in one dwelling unit throughout their lives, accentuated spatial specialization could end up being highly inappropriate for future edifices unless portability capabilities are also incorporated in them.

Consumerism is another issue likely to continue its fundamental influence on the evolution of American habitats. The continued rise of a consumer-oriented society in this nation would probably result in higher goods turnover, throwaway policies (more or less affected by spot resources availability) and increased building design stereotyping and prototyping. It would also contribute to an increased loss of privacy and to the introduction of still more "created" social needs.

Increased consumerism, could also facilitate the creation of large corporate conglomerates that could end up taking over many aspects of social life-styles, a trend which would facilitate the development of the mega-builder corporate structure concept.

In many ways, North America has an invaluable tool in its consumer-oriented society, since economic, political, cultural, technological or natural crises can be much more readily handled, or even circumvented, through adequate public responses. The reaction to the energy crisis of the last few years is a clear example of this. Thus the growing forces of consumerism should not be viewed entirely as negative social trends but as phenomena whose effect on the future built environments of this nation will depend solely on the relative role assigned to consumerism by society in the years ahead.

The probable influences of life-styles on the evolution of American shelters, therefore, can be defined in terms of three important issues. First is the possibility of increased leisure/entertainment type activities as well as full- or part-time work-at-home patterns for certain occupations in the coming years.

Second is the need to incorporate in future shelters a new "mobility" concept so that the transitorial living patterns of the dwellers can be properly supported by their habitats.

And third, since consumerism is a direct product of economic conditions, resources availability and technological developments, it can become an extremely valuable aid towards the attainment of spatial adequacy and environmental safeguards in future urban and suburban built environments.

SOCIOPOLITICAL ISSUES

Among the most significant determinants of the future North American edifices are those involving government growth. This is a complex area of study because, although almost everyone agrees that American government has gotten too big, wasteful and inefficient, as the areas of human endeavor expand, regulatory action must grow to keep pace with them. If one eliminated automobiles from the American scene one would also eliminate a myriad of regulations, and if one eliminated television, radio and public education, one would end up reducing the size of government—but at what price?

Governmental intervention in the evolution of built environments has traditionally impacted on their cost. Rarely will government interference reduce prices. Increased regulations and safety measures for nuclear power plants, for example, will increase the cost of electric power, just as further regulatory interference in building construction will increase the cost of shelters. Unfortunately, there are signs that in years ahead there will be increased government influence on future habitats.

Reduction of government interference in the evolution of habitats would allow for natural supply/demand solutions to most problems and would shift a lot of the responsibilities to local communities and neighborhoods, thus allowing adequate regional readjustments. Continued government growth, on the other hand, could cause the rise of special interest groups, higher taxation rates, and increased governmental control of economies at all levels. This would result in accentuating the loss of privacy, reducing free enterprise, and further damaging the nation's competitiveness in international construction marketplaces, a trend which has been affecting the American building industry for well over a decade. Since 1975, the position held by the United States in the international construction standings, for example, has dropped from third place (behind West Germany and Japan) to twelfth place, and the total share of construction firms on the international marketplace dropped from 10.3 to only 1.6 from June 1975 to June 1979 in the Middle East alone.[2]

On a national basis, increased government interferences in the development of habitats would continue to drive up the cost of building construction. The National Association of Homebuilders estimates that government building standards aimed at saving energy would increase the average price of a new home by $1,755.[3] Government interference in building construction, however, is one trend which may be hard to avoid, since many new regulatory measures will be the result of new practices brought about by progress within the industry. Thus it is very likely that as new technologies are developed, additional regulations affecting systems and products will also come into being. How drastically these expanded regulatory forces will affect present trajectories is hard to predict. However, one fact does remain clear: the overburdening role presently played by government regulations in American habitats is unlikely to change before the turn of the century. Not only are the leaders of the 21st century being raised and educated under present conditions, but also a great majority of the buildings that will exist then have already been built or are being planned within the present network of regulatory constraints.

The laudable efforts to reduce government growth in this country may also be imperiled by other issues which demand more rather than less regulatory interference in sociological interactions.

Crime, for example, is an issue whose importance is unlikely to decline. Lower crime rates would mean a reduction in security features and costs for all American habitats, thus facilitating shelter maintenance, operation and insurance. It would also permit optimum use of parks, public urban spaces and green belts; unlimited possibilities for urban renewal projects; better planned urban and suburban developments; safe and efficient implementation and use of mass transit systems; better shopping, living and entertainment schedules; and reduced investment and maintenance costs for law enforcement, judicial and detention facilities. To these benefits one must add the unlimited possibilities for new construction materials whose cost effectiveness is only challenged by their limited resistance to defacement by crime and vandalism, such as paper derivatives, plastics and treated softwoods.

A significant reduction in crime rates, however, is highly unlikely to occur in the near future, and this could constitute one of the most formidable brakes to the progressive development of American society and its urban and suburban built environments.

Continued high crime rates would further accentuate urban/suburban decay throughout the country. Under such circumstances, urban renewal projects would become mere palliatives for an overall situation way beyond the control of anyone.

Increased crime rates would also provoke stricter police measures and thus make shelters more vulnerable to foreign intrusions by either criminals or crime fighters. Unless crime rates are reduced, buildings will continue to be defaced, their maintenance costs will soar, and the appearance of many urban sectors will inevitably deteriorate. This would result in more "walled-in" or clustered building arrangements, private neighborhood type developments and in the proliferation of mini-communities within guarded perimeters, thus promoting increased urban fragmentation and suburban sectorization.

Presently the indications are that for the remainder of this millennium the trend will be towards an increase in security measures in most American habitats in the form of locks, alarms, fingerprint and voice recognition systems, weapons, guards, neighborhood crime watch programs, entrapment systems, and even some radical changes in the present ease of accessibility to most dwellings.

Unquestionably, crime does constitute an area of serious concern in the sociopolitical development of this nation. But American society is also confronted with another issue of even greater significance: the gradual social fragmentation caused by an increasing diversification of special interest groups. With a greater proliferation of activities, occupations and interests, and a larger population base, the unification of individuals within special interest groups designed to add more weight to their causes is a logical consequence.

This social trend could eventually facilitate further government growth and increased regulatory interference on land usage, habitats and real estate affairs. In this regard, this nation risks creating power elites with near-absolute control over the future evolution of American edifices.

The fragmentation into narrowly defined sociopolitical groups would have the positive effect of preventing the total control of most physical environments by large corporate structures, but at the same time, it could tend to slow down the progress and economic growth of many regions and, conceivably, of the nation as a whole.

This brief analysis of sociopolitical issues, therefore, underlines four major areas of concern. The first is that government will most likely continue to expand its sphere of influence as the areas of human endeavor proliferate, and this would constitute a setback to the prompt resolution of many problems presently faced by American habitats.

Second, crime could become a much more serious problem in the years ahead, and the sociological implications of the eradication of crime could be one of the most dramatic issues to confront industrialized societies in the future since crime is undoubtedly one of the issues most likely to replace cancer as the number one social concern of the next millennium.

The third point is that continued urban and suburban decay during the next few decades will eventually force massive reconstruction programs aimed at salvaging the investments already made in American built environments.

And fourth is the general conclusion that while forecasters inherently tend to become optimistic about the future when dealing with resources availability and scientific possibilities, a hard look at some of the sociological issues presently dominating the American scene demonstrates that the outlook is not really that promising.

It is in this lack of correspondence between its sociopolitical characteristics and its scientific advances that the industrial society fails. Obviously, the attempts to compensate for this ever-widening gap cannot succeed in the short course of one generation. There are many cultural factors involved that make the transition from one stage of the American quest for sociopolitical improvement to the next a complex and difficult one.

CULTURAL FEATURES

One significant change in the cultural fabric of the United States would be an acceleration in the family fragmentation patterns which have characterized this society since the early years of the Industrial Revolution. If a gradual fragmentation of the traditional family unit as it is understood today were to continue, it is very likely that a new concept of a familial conclave—probably based on specific household structures—would emerge as the American "family" of the future.

The marked individualism and fast pace of modern times could reinforce this trend despite the possibility of a larger percentage of the population spending more time at home, since it would take more than work-at-home activities and increased leisure/entertainment time to form and maintain cohesive family units. Moreover, increased work-at-home patterns could cause more complex and varied household structures based not only on emotional and blood ties but on business, educational and economic arrangements as well.

In addition, further family fragmentation would accentuate social isolationism and loss of privacy, and would lead to smaller households and to a significant increase in smaller multifamily type dwellings, as well as increase population mobility and produce more complex and diverse kinds of households. The increasing number of household types, for example, is a trend that has intensified significantly during the last decade. As early as 1977, a survey conducted in a poor sector of Chicago encountered no fewer than 86 different household structures.[4]

What this nation may well continue to experience is not a drastic change of traditional family arrangements, but a modification of its traditional concept of "household." In practice, however, some of these modified household/family units could function in a manner not unlike that of the traditional households of pre-Industrial Revolution years, with common interests, activities and work and study patterns.

It is noteworthy that, because of their levels of education and communication and their climatic conditions, the regions having life-styles most likely to adapt well to increased at-home activities and extended household/family concepts would be precisely those areas which many forecasters visualize as decaying national regions: the northeastern and midwestern states. Here again is another indication that in the 21st century those regions could be thriving, carried by a wave of novel developments and evolutionary occupational patterns into economic prosperity.

Another significant cultural development in the coming decades would be an increase in American sexual egalitarianism. This could contribute to a better qualified work force, but it could also cause further changes in traditional family structures. Actually, in the not too distant future many of those altered living arrangements could start producing some radical variations in the floor plans of shelters, since functions like cooking and laundering may be done indiscriminately by any household member at any particular time, or even simultaneously by several family members. It is also very probable that within a few years robotics and automation will provide American habitats with a whole new set of options for executing many of today's most time-consuming household chores.

Another important cultural development would be the fragmentation of age groupings. This could possibly redefine the relative social and economic status of age brackets, as well as their interactions. Widened gaps between age groups would increase social consumerism and youth-oriented cultural trends, possibly accentuating features in bathrooms or "health rooms" aimed at complex make-up activities, facial treatments, exercise routines, etc. Age-related social fragmentations could result in discriminatory attitudes towards older and younger Americans alike for shelter acquisition or rental contracts. And this would in turn impact urban and suburban evolutionary patterns, and population shifts throughout the nation.

A likely—and desirable—development, however, and one that would eventually reunite younger and older Americans under a more comprehensive social umbrella, is extended life expectancy. This is possibly the single most important factor that could contribute to bridging age gaps in this nation. It is no wonder that 1980 saw the election of the oldest man ever to be chosen as president of the United States; in fact, that is one record that will probably be broken time after time in years to come.

The advantages of increased life expectancy are factual and numerous: an improved work force, more even population distributions throughout the urban/suburban fabric, reduced crime rates and improved household/family concepts. It would also be very likely that the future could bring a significant increase in "inter-age group" marriages and other relationships as life expectancy is extended, thus impacting on the financial structure of habitats.

Closely linked to traditionalism are the paths of development likely to be followed by morality, religious beliefs, etc. In future years it would not be surprising to find religious services being conducted mainly through television and videotex units, thus making their social outreach even more widespread than it is today.

Ethnicism, however, is one issue which may not be easily dealt with by the American society in the coming years. In this respect the only logical conclusion is that the key issues behind most of the racial problems faced by this country are very unlikely to disappear for at least a couple of generations. During the next few decades the United States could easily continue to experience occupational and income level fragmentations based on racial origins, increased governmental interference in matters dealing with racial discrimination, whether for housing, urban patterns or land development. Further accentuation of economic conditions or social status based on ethnic groupings could also set the stage for more racial conflicts within urban/suburban neighborhoods, contribute to the reduced economic growth in some counties, lower educational levels in specific urban sectors and cause increasingly complex suburban evolutions.

Ethnicism, however, is a phenomenon that is likely to diminish in this country in the years ahead, mostly because of its damaging socioeconomic effects. If a significant improvement in this area were attained, future American generations would experience much better regional distributions, and many of the urban and suburban problems that plague American cities would be resolved, resulting in improved work forces, reduced crime rates, and much better organized developments of urban/suburban habitats.

Summarizing, we can conclude that a general overview of the cultural features of modern American society suggests four basic areas of interest relative to the future development of edifices in the United States. First, the traditional "family" structure will probably change into an expanded household arrangement where increased home activities and improved communication patterns may facilitate new and highly sophisticated living arrangements, based not only on familial ties but also on varied skills and interests.

Second, because of these basic changes in household and family structures, products, inventions and appliances aimed at saving time or facilitating the execution of household chores are likely to be widely accepted throughout the country in the coming decades.

Third, although the traditional gaps between young and old may not change drastically, the relative age brackets at which these gaps may occur could be bridged due to the possibility of increased lifespans in years to come.

And fourth, ethnicism is a social characteristic that will probably diminish in importance in future years since its political basis and socioeconomic implications are contrary to most guidelines presently followed by those most likely to shape the future of this nation.

14
Guidelines

When the issues analyzed in the previous chapters are considered in the function of habitats, several trends come to light, many of which involve factors that are already affecting American built environments.

A general overview of these basic trends can be summarized as follows:

- The United States population growth during the next decades will be mostly due to two major factors: foreign immigration and the increased life span of American citizens. According to the Bureau of the Census, the number of persons of Hispanic origin (the largest migratory inflow) rose 2 percent in the last decade, adding up to a total of more than 14.5 million people.[1] And the number of people living alone in the United States jumped 66 percent in the last 10 years—a total of about 6.4 million people[2]—of which the largest number were elderly women, more than half of them widows over 65 years of age.

  It is unlikely, however, that these population increases will be evenly distributed. Presently, the total number of states showing population inflow within immigration patterns exceeding 2 percent (all in the Northwest, Southwest, and South) triple those showing population outflow (all in the Northeast and Midwest).[3] And it has been forecast that, at present migratory rates, by 1990 an average of 3 out of every 4 people in this country will try to settle in coastal regions.[4] Although these forecasts may be somewhat exaggerated, these migratory trends may continue during the next two decades or so, but once a balance is gradually established the distribution pattern would be likely to shift its course towards those areas with present migratory outflow and stabilized population.

Moreover, the specific types and characteristics of the immigrants this country might expect until the turn of the century would be mostly Hispanic and urban and would tend to concentrate in the southern and southwestern regions, while national population shifts could continue to drive other Americans westward. In each case the characteristics of the migratory groups would tend to seek and permit the urbanization of rural areas in these regions.

The belief that increased migratory outflows will cause eventual decline in regions with stable or declining population indexes is unjustified, since larger population concentrations in growing areas and the substantial capital already invested in those stabilized regions for habitats, utilities, communication and services could equalize—and eventually maybe even reverse—these trends.

- Resources availability could play a key role in the evolution of this nation's habitats by forcing spatial modifications in edifices, and by introducing new building types, construction systems and materials substitutes.

Of all available resources, however, land is the only one for which there are no possible "undiscovered" reserves. An extremely important lesson to be learned from the awareness of limited resources is the importance of keeping urban/suburban growth from seriously damaging arable land, forests, national parklands and other irreplaceable acreage.

In the next millennium, agricultural, recreational and forested land could be seriously encroached upon by unchecked urban/suburban horizontal growth. Because of this, finding alternative means of expansion (vertical growth, underground dwellings), considering possible substitutes (marine structures, outer space colonies), and planning for the adequate development of utilities and services to support these innovative life-styles will gradually become more imperative.

Meanwhile, as the affordable acreage dwindles, lot sizes will probably continue to decrease, utility costs will continue to rise and living units may tend to get progressively smaller.

- There will probably be few drastic modifications in the general evolutionary patterns of shelters during the next two decades. However, some changing characteristics could become more apparent as changes in life-styles initiate new trends in habitation, such as reduced operational parameters, emphasis on spatial efficiency and resources conservation, new financing arrangements, greater emphasis on research and development, and the increased growth of systems that facilitate user-built techniques and building equipment maintenance. On the other hand, there could also be greater governmental and regulatory interference, more investor-controlled building projects and more complex owner/user arrangements.

A desirable future development within the construction industry would be to free workers from its labor-intensive processes so that they could be transferred to other expanding areas such as agriculture, land mining, oceanic farming and mining operations, and outer space explorations.

Reducing the high cost of utilities will probably be another major concern through the next millennium. This could be accomplished by the development of more efficiently designed and operated appliances, water efficient plumbing systems, active and passive solar technologies and products, adequate energy conservation measures, and the gradual automation of shelters. Improvements in the materials and construction systems presently employed in the execution of edifices would also alleviate this problem.

In the next decade, passive and active solar technologies are likely to dominate the development of most buildings in America, and during the early years of the 21st century underground construction could become an increasingly considered alternative. Private automobile usage is not likely to drop significantly in the near future, but transportation in the next millennium could experience a gradual decline of privately owned vehicles in favor of improved mass transit systems and automated transportation.

Conversely, the rate of progress that would facilitate many of these developments could be continuously slowed down by excessive government growth, high crime rates and sociopolitical fragmentations. Nevertheless, factors such as reduced resources availability linked to population growth, new scientific advances, economic pressures, occupational parameters and changing life-styles could become major determinants that would eventually force many of those spatial and transportation improvements to take place.

- In the years ahead the major sociopolitical influence on habitats in this country could probably be on the improvement of decaying urban/suburban sectors. Aside from being socially undesirable, continued urban decay beyond the 20th century would be economically unaffordable. The first steps towards confronting this fact may already have been taken.

  Crime, however, is an issue that cannot be disregarded when considering urban and suburban transformations. If crime is not adequately controlled, it will become one of the most significant stumbling blocks in the path of urban revitalization.

  Barring the negative influences of high crime rates, one could look ahead to such positive urban developments as the eventual transformation of many shopping/entertainment centers into self-contained commercial mini-cities (domed or underground complexes), which could end up incorporating short-term stay hotel/motel facilities designed to house patrons overnight as workweeks are reduced, leisure time increases and schedules become more flexible. In effect, what the emerging cities of the next millennium could easily consist of is a combination of modern centers of the type just described with older, more traditional types of structures, making up urban complexes which could accommodate the changing life-style patterns of American society while still maintaining the essential characteristics of *dated* cities.

  Urban and suburban sectors could also be further transformed by a continued growth of transportation and communication arteries paralleling major interstate highways. This trend would facilitate the booming growth of areas located at strategic points between major urban nuclei, such as Houston-Phoenix-Los Angeles, Chicago-New York, and St. Louis-Houston.

- Traditionally, the wealth and effort allocated to the development of shelters have been among one of the highest expenditures borne by man. The future evolution of habitats should be aimed at reducing the relative cost of the development and maintenance of edifices as a percentage of personal income. In order to achieve this, two things would have to happen: the first would be the development of construction systems and materials that would permit savings in the labor-intensive processes of building and maintaining edifices; and the second would be the creation of shelters that emphasize optimum space utilization while still maintaining their functional adequacy. Neither trend is likely to develop significantly in the near future. However, the proliferation of some new building types (high technology centers and modern office or entertainment complexes) may help initiate some of these construction characteristics and spatial redefinitions.

The effects of these changes would ultimately be reflected in the relative cost outlays for building construction. However, in this light some building types offer much better opportunities for cost effectiveness than others: condominiums and other collectively owned and maintained dwellings (whether for housing or commercial property) would be much more prone to cost benefits than single-purpose/individually owned structures. Glass covered greenhouse-type gardens, for example, can provide indoor/outdoor benefits to 30- or 40-story skyscrapers or to underground structures as adequately and efficiently as to one-story structures.

To the advantages of the multi-user type building one must add the benefits of shared rental structures, varied options to rent or buy, and other financial arrangements that could become even more advantageous in cases where mobility was a factor, and could also become very appealing to large corporations which might end up investing or speculating in housing as an employee fringe benefit, or as a profitable side business.

• It is unquestionable that the North American housing unit would be significantly affected by life-style changes.

The American home of the future could be altered to accommodate increased work, education and entertainment activities that probably could not be carried out within the spatial parameters of many existing housing units.

The trend towards increases in work, entertainment and education activities at home makes a lot of economic and social sense because the average modern shelter is an extremely expensive commodity to rent or purchase and maintain, and its potential usage is wasted for 35–40 percent of the time because its dwellers are either working, studying or entertaining themselves elsewhere.

For certain occupations, shortened workweeks or increased work-at-home patterns could consequently trigger changes in the present patterns of the use of some commercial types of edifices such as office buildings and professional services complexes. Thus many of these structures may end up being remodeled to incorporate new functions or simply being razed to accommodate new building types.

- Among the building types most likely to proliferate in the near future are: entertainment and sport centers, computer retail stores, high technology manufacturing buildings, communication centers, research centers and laboratories, repair and maintenance shops, airport/heliport terminals, manufacturing and industrial buildings in general, home improvement centers, health and physical education complexes, recycling plants, specialized training centers, nursing homes, and special health care facilities and hospitals.

Among the building types whose function is likely to change somewhat one could cite: banks, libraries, professional service buildings, office buildings, art galleries, wholesale distribution centers, insurance/credit union buildings and schools in general. And among those likely to decrease in use could be: parking ramps, cemeteries, gas stations, banks, churches, specialty shops, elementary and secondary schools, small retail and commercial centers, movie theaters, large office complexes, government complexes, post office buildings, dump yards, and military installations.

The social factors likely to influence the evolution of habitats in the United States indicate a gradual need for the development of new habitable environments. The population characteristics, for example, point to an increase in the number of households and a decrease in the number of household members. This will demand more spatial flexibility, and thus facilitate the development of spatial prototypes. More diverse household types do not necessarily mean more diverse spatial arrangements, since the latter are defined by functions and not by human beings. Thus it is from the diversification of functions (one of the essential proponents of spatial prototyping) that the variety of spatial arrangements will develop. Moreover, issues such as crime and governmental interference may further stimulate the development of serialized or standardized space/use envelopes.

The forces that will define the rate of development of future shelters, on the other hand, will be greatly affected by the availability of land, and the ratio of land cost versus building cost will be likely to increase in the decades ahead. Thus a reduction in the overall cost of

habitation can only be achieved through those alternatives which reduce the relative costs in the shelters themselves, or those which would
increase their time-utilization factor.

The problem of land development as a percentage of overall property cost is compounded by the fact that traditionally the land selected
for habitation has been the easiest one to develop and thus the least
costly. The situation in which the relative cost of shelters is to be further reduced to facilitate their affordability is one that particularly calls
for prototyping and serializing edifices, since mass producing large
components or even complete habitation units is one sure way of decreasing costs.

There is another approach, however, to the relative discrepancy in
the land/housing cost ratio and its increase as a percentage of personal
income, and that is to let that increase be justified by the fact that more
time would be spent in dwelling units, so that the use/cost ratio of
housing would remain relatively unchanged. In fact, more time spent
at home (resulting from increased work or study activities carried out
within the home) would translate into more time available for home
care and maintenance, since much of the time presently spent away
from home is, in effect, wasted on transportation and other low productivity tasks.

If an equal ratio of cost of shelter to income per capita were maintained, the final result would be an actual reduction in the relative
cost/use ratio of the dwelling. Also, other factors may contribute to
ease the higher cost of habitation: if an employee were credited with a
mere 30 to 40 percent of his/her share of the overhead costs borne by
an employer in the form of special subsidies for working at home even
on a part-time basis, depending on the circumstances the employee might
be able to afford to pay double for his/her housing costs.

The physical and economic impacts of shelter development, prototyping, land availability, urban and suburban growth and housing values, however, will ultimately be a direct function of the interaction of
resources and technology. Here the possibilities fall within two broad
categories: alternative placements and operational systems.

In the first category, the possibilities are obviously limited by the
physical landscape: habitats may develop sideways (eliminating side
yards and backyards) or upwards—a common development with great
limitations—both of which would increase overcrowding in existing

urbanized sectors and cause heavy strains on existing utilities and services, unless a progressive independence from the latter could be achieved. Habitats may further develop into presently uninhabited regions having conditions hostile to life, but in doing so they could create situations which would pose serious environmental and economic problems. They may develop underground, above or under water, or they may even take to airborne structures or outer space colonies.

In the second category of possibilities, that of operational systems, the basic elements are: labor (likely to be reduced through the introduction of automation and robotics in the building processes), materials, the reduction in size of built environments and land allocations, the adequate incorporation of new technologies, the reduction of buildings' operational and maintenance costs and the probable achievement of almost complete independence from cumbersome utility services, and advances in standardization and recycling. In fact, it would not be surprising to find the durability concept presently attached to shelters modified to cover only 50 to 60 percent of their contemporary life spans. This issue is very closely related to standardization, recycling, and relative costs as well as to the growth and change characteristics that will play so vital a role in the future evolution of American edifices.

The interaction of many of these issues could eventually lead to other developments: more free time may mean more time available to maintain a less costly habitable unit having a shorter life span. The use of automation and robotics to further help with these tasks could easily redefine many of the conceptual approaches to shelters' development, care and ownership.

Many of these changing characteristics and the functional and operational redefinitions to be expected in years to come can be foreseen upon consideration of the issues previously mentioned. During the years of the Industrial Revolution, edifices (especially houses) were redefined as "machines to live in." This definition represented the new social adaptation and integration of machinism into everyday life. Today that definition could be modified to read "automated dwelling assemblages," thus incorporating the electronic age.

This redefinition would presume further changes in the traditional characteristics of edifices as well as the introduction of new ones. Among the redefined values embodied in the human shelter one will be likely to find the following:

1. New concepts of functional and spatial adequacy
2. Controlled environment
3. Self-sufficient operational and maintenance goals
4. Complex security/protection/privacy features
5. Improved health/hygiene/comfort features
6. Improved communication systems
7. Redefined durability requirements
8. Social identification and traditionalism

And among the new characteristics most likely to be imposed by future socioeconomic and technological developments one could find:

1. Work/education/entertainment capabilities
2. Mobility/portability
3. Climatological and topographical adaptability
4. Expandability and contractability

One of the newly defined goals that American habitats must meet during the next millennium is the adaptation to the increasingly complex mobility patterns of American life-styles. This could be achieved by greater emphasis on current "buy and sell" arrangements, by the development of demountable/mobile or relocatable types of shelters with land-lease contracts built around them, or maybe even by radical departures from traditional ownership into mere "occupancy" concepts, where corporations or housing companies satisfy the habitation needs of the American people as a rental/leasing service, as a negotiable commodity, or even as a fringe benefit.

The need for both mobility and permanence seems to be one of the most significant paradoxes presently confronting American habitats. But this is only one issue; as we have seen, the requirements imposed by other aspects of changing life-styles (work and entertainment patterns for instance) will also be key factors in the future redefinition of this nation's urban and suburban landscapes.

The many tools and procedures by which some of these changes could be realized, as well as the entities into which American shelters may evolve, are the subjects of the next three chapters.

15
Habitats: New Horizons

In general, forecasts fall into three basic groupings: probabilities, possibilities and long-range visions. In studies of the future, all logical developments should be considered but priorities are defined in sequences that begin with probabilities and end with long-range visions.

It is highly unlikely that all of the possible transformations of American habitats mentioned in the following pages will come to pass, but it is almost certain that from among these possibilities a general sketch of this nation's future edifices will emerge.

To present these possible developments, this chapter will be divided into two major sections: one dealing with the spatial components, individual characteristics and functions of habitats, and the other focusing on the possible futures of communities, cities and communication.

COMPONENTS, CHARACTERISTICS AND FUNCTIONS

It would be impossible to include here a study of *all* the spatial components of edifices, but a general summary of some of the most essential ones will serve to illustrate the trends likely to be followed by many of them.

Vestibules and other entryways are probably the most important organizational/distributional spaces in many buildings and are likely to retain (and even increase) their importance in residential units if work-at-home patterns become more prevalent in future years, and if functional fragmentations within the North American built environment continue to multiply. In those single family dwellings incorporating work and living spaces, vestibules and entryways could be used to define or isolate work and living areas some of the time while at other times they could easily double up as auxiliary living or entertaining spaces. And in large building complexes, robotics and automation could also transform vestibules into actual "transportation modules"—mobile entryways that could transport visitors to their destination.

Living rooms and family rooms are very likely to blend into a single area adequately zoned to encompass the functions of both spaces. In fact, it is possible that "great room" designs (a concept similar to colonial dwelling arrangements with most household functions taking place within a common area) may influence the layout of many future American housing plans, so that the simple incorporation of movable, demountable or sliding partitions would permit the definition and rearrangements of zones within their envelopes. Variations of this concept have already become popular in commercial, public and industrial buildings.

There are some areas that, by nature, are not easily modified to incorporate dual spatial functions, but that may be simplified in the future. Kitchens—having consistently expanded their size as economic and technological affluence have increased—may become simple "nutrition centers" if the development of American food processing systems, automation and robotics combine to simplify many of the burdensome tasks presently carried out in modern kitchens.

Bathrooms, on the other hand, are areas likely to evolve into complex health and hygiene support units. "Living bathroom units" (concepts that encourage common use by all household members), new bathtub, lavatory and water closet designs, and even waterless type toilet fixtures could easily trigger significant bathroom transformations. Recent publications, for example, have noted several deficiencies present in most modern bathroom and toilet fixture designs (such as poor and cramped spatial layouts, the inadequate ergonomics of bathtubs, the inconvenient positioning of faucets, the unsanitary characteristics of water closets, and the inefficient mounting heights and proportions of modern lavatories), and have proposed drastic modifications that may end up revolutionizing bathroom designs in future years.[1] Also, since the mid–70's combination sauna/shower units that provide full environmental control (from ocean beaches and tropical rain forests to arid desert climates) with stereo tape player, artwork options, and other amenities have been commercially available.[2] Moreover, if, as some have ventured, some of the present cleansing and sterilization processes could eventually be achieved through the use of special gases or lighting systems, the functions of the bathroom would change drastically.

Storage is one spatial component of American edifices (an obvious by-product of consumerism) whose evolution could be drastically affected by changing life-styles. Americans visiting abroad are continuously amazed at the small amount of storage provided by most foreign dwellings. The fact is that, throughout the world, people own and store much less than Americans do. The cost of owning things does not stop with their purchase, it translates into square footage of storage space. For decades now, the average American shelter has been increasingly saturated with storage space that, paradoxically, always seems insufficient. In an average 1980 3 bedroom, 2 bath home of approximately 2,060 square feet, it is standard to find anywhere from 180 to 200 square feet dedicated solely to storage.

It is unlikely that this pattern will continue indefinitely. In future years the American shelter could experience a gradual reduction in its overall storage, a rising demand for "built-in" features and efficiently designed and used storage space. This trend, which would be facilitated by automation, advances in food processing systems, improvements in textiles and modifications of packaging and disposable systems, could be a direct result of the probable reduction in the overall size of most shelters.

Bedrooms could also change radically in the future. As shelters diminish in size, the spaces assigned to bed storage may end up having more than one function as other furniture items are included in them, with water or air-filled bedding systems further altering their layout. Beds may eventually be stored vertically or evolve into actual "sleep capsules" which could be stored under floors or above ceilings and be raised or lowered only during sleeping hours. These "sleep capsules" could provide complete environmental, sleep/awakening and therapeutic controls, and also monitor the physical activities of the user much like intensive care units in modern hospitals. Thus with flexible bed and clothing storage, the functions of the traditional bedroom could easily expand to include those of a study, workshop, media room or even greenhouse in future years.

Garages are another element that may change if communication patterns and mass transit systems improve. Moreover, increased work-at-home patterns for certain occupations may not necessarily entail "fixed" spatial work areas. Many future "work modules" could become mobile units that "plug" into dwellings for indefinite periods of time, but that also have the capabilities of relocating and transporting their occupants as well. "Work shuttles" (capable of land, water or even air transportation) complete with computers, telephones, etc.—a relocatable "mini-office" of the future—could very well constitute the second phase of the transition to widespread mobile work patterns. Eventually large work, shuttle-parking, or gathering areas (for temporary/semi-permanent occupancy) could replace the large office complexes of today.

Gardens (indoor or outdoor) may also undergo radical changes. In all likelihood, the size of most private outdoor spaces may have to be reduced in the coming decades, thus making their effective use much more critical. Several forecasts have contemplated the use of private vegetable gardening and home food production as a possibility for future years. However, there are serious problems with this type of agricultural decentralization: first, there are some innate limitations to widespread food production processes (like the individual capabilities of developing adequate agricultural tasks); and second, even if automated processes and advanced robotics were specifically developed for these purposes, total decentralization has never been proven an effective agricultural production system. Further advances in vegetable gardening, hydroponics and animal farming *may* alleviate future food crises, but to expect technoscientific advances that would eventually justify inefficient methods of agricultural production is to project too far into the future. It is possible, however, that plantscaping may be of increasing importance in the future evolution of American built environments.

Among the technoscientific advances that could facilitate many of the developments mentioned above are the automation of appliances and other household features; the introduction of robotics for household service, maintenance and repair functions; and the development of computerized or fully automated dwelling units. Although still in the experimental stages, computers capable of managing entire shelters have been developed already, and existing technologies that combine electronic devices with telecommunications can readily control access, alarm and emergency calls, temperature and noise levels, solar effects and ventilation, lighting and even lawnmowing.[3]

Also in the process of development are commercial, industrial and public buildings that have built-in sensors (a technology commonly used by satellites in space) that feed information to a central computer which can balance their mechanical and electrical systems and even reorient the structure, should it become necessary.[4]

Further developments of passive and active solar technologies, re-cycling of solid wastes and gray water, self-cleaning building surfaces and rooms,[5] and specially designed "mood altering" habitats are other recent innovations that may have an impact on future edifices, while three-dimensional photography and holography combined with the fur-ther influence of television on American dwelling units are also likely to modify modern home entertainment and leisure time patterns in fu-ture years. In this respect, consideration has already been given to the creation of rooms totally surrounded by media screens, thus actually producing three-dimensional "video environments."[6]

In general, the square foot per occupant ratio of American dwellings is likely to decrease in the decades ahead. Recent forecasts have ar-gued that in the near future it may even take two household units to be able to afford a private residence which would probably be de-signed or remodeled to accommodate the obvious living pattern vari-ations.[7] Other reliable forecasts maintain that the average shelters most likely to be built within the next two decades will be smaller than the present ones and have fewer amenities.[8] Likewise, more efficient building types such as town houses, multifamily dwellings and two- or three-story houses could constitute the bulk of the residential building marketplace in future years. Not all specialists agree with the forecasts that call for explosive shared and multifamily habitation patterns, but there is one common issue on which they *do* agree: in the near future, American shelters will be getting smaller. And this is especially true of single family dwellings, which depend heavily on available land, plentiful materials, cheap energy and the long-term commitment of large capital expenditures.

It has been calculated that at present rates, the average price of a new house in this country—a $93,000 cost in mid–1983[9]—could exceed $150,000 by 1990.[10] At these levels, over half the income of an average dweller could easily be spent solely on housing costs. Conversely, the size of residential lots has already begun to decrease gradually from its 1979 peak of 1,190 square meters per lot, signaling a trend similar to that of housing units. A recent study showed that in the 1970's the total area for houses built ranged from 2,000 to 2,500 square feet. However, by August of 1982 a survey of approximately 200 builders conducted by the National Association of Homebuilders found that about 60 percent were building housing units that only ranged between 850 and 1,400 square feet.[11] It would not be surprising to find spatial parameters of present housing reduced even further in the coming years, to sizes even smaller than those of the 50's and early 60's. This may be a positive trend, since at its 1980 average of approximately 870 square feet per person (in excess of two rooms per dweller), the typical American housing unit had become, in fact, a very wasteful entity.

Other possible changes and adjustments in American habitats to accommodate larger concentrations of urban dwellers may include the clustering of residential units or secondary housing units on larger lots, as well as the partial or complete elimination of side, front or rear yards.

Modular housing units will also play an important role according to most forecasts regarding future habitation. The concept of modular construction is not a recent one. For decades now, offices, schools, factories, homes and even commercial establishments have been built using this building approach, which developed mainly out of strategic war needs. The theoretical advantages of this building system are numerous: greater quality control, spatial flexibility to accommodate growth and change, possibilities of building stacking or clustering and, above all, the capability of developing undifferentiated single *or* multifamily dwelling edifices with simple and economical predefined spatial components. Several modular building systems have already been developed, some resembling concrete boxes that can be stacked, jacked up, plugged in, or slid into steel or concrete frameworks[12] with their utility connections finished on site (some of which, by the way, may not be necessary at all in the future), and offering the possibility of locating doors, windows or entire living components according to the most desirable orientation. Also, small prefabricated low cost housing that varies from 480 single story units to 850 two story houses designed to grow through room additions are presently being developed in Third World countries.

Most of the problems faced by modular construction systems today involve the transportation of components. However, these limitations could easily be circumvented in the future by new transportation channels (such as air and water) or improved transportation systems, as well as by the use of specialized vehicles combined with mobile fabrication and assemblage techniques.

Advances in these areas could lead to the widespread development of demountable or reusable edifices and even to the proliferation of entirely movable dwellings. New concepts in this area have already been developed by creating dome-like circular structures with rotating walls that define spaces within a "great room" plan concept. Relatively lightweight (made out of aluminum), the exterior finishes offer all the variations of baked-on enamel colors while offering the durability and weather resistance of a long-tested construction material. The original models offer up to three bedrooms and two baths. Other modules can be added as necessary (in almost any direction) to accommodate growth, and the developers maintain that theoretically, entire communities of these modules could be developed by joining (or enclosing) clusters of four or more units at a time.[13]

Combined with existing dwellings—as rental units, expansion cottages, work modules—these modular shelters could easily become the much needed mobile house of future years. These developments would, in turn, reinforce the concepts that call for increased stylistic unification and prototyping of future habitats. Factory mass produced models offer little choice for significant formal variation, except for superfluous details. Individuality in future shelters may be increasingly focused on internal rather than external characteristics.

Other habitation alternatives for future years that have been seriously considered recently are the development of earth-sheltered structures and underground buildings. The concepts are not new. The first American earth-sheltered buildings were developed by the colonists in the form of dugouts on hillsides for temporary dwellings over 300 years ago. In fact, the concepts of earth-sheltered and underground dwellings date as far back as prehistoric times.

The advantages of underground structures are many: noise and vibration control, heating and cooling savings, environmental controls (wildlife habitats, rainwater runoff), organic waste recycling and longer lasting structures. Other significant savings relative to building features include: foundation costs, weather protection and durability, maintenance-free exteriors, noise control, and excellent fire resistance characteristics.

Of all of these, the most significant advantage of these types of shelters is their energy efficiency. An underground building only needs enough energy to raise the temperature from a constant 54° F to 72° F. The National Bureau of Standards estimates that 100 billion dollars in energy could be saved over the next 25 years on residential units alone if the heat transmission characteristics of American housing units could match those of underground structures.[14]

There is also the advantage of the minimal environmental impact posed by underground building structures. Natural lighting can be achieved by the proper use of light wells and skylights, and even the actual reflection of exterior landscapes could be transmitted to deep underground levels by the incorporation of new optical systems within the buildings' envelopes.[15]

Underground edifices would be more adequate than most above-ground structures in regions with extreme climatic conditions. The likelihood of their proliferation, however, depends on the future advances of excavation techniques (several rock melting and shattering techniques, nuclear powered probes, and laser beams have been proposed)[16] and on the public acceptance of underground living (which some studies indicate may be on the rise). The negative aspects of underground habitats, however, are their spatial rigidity (growth would be difficult to accommodate) and their lack of mobility. There are also complex psychological ramifications. The full effects of permanent underground dwelling (over 80 percent of the occupancy time) have not been properly studied or documented.

Earth-sheltered structures have many of the advantages cited previously, but they are a lot easier to develop than fully underground structures. Their adaptability to hillsides and mountainous terrains may contribute to their popularization in future years.

Marine structures have also been the subject of many recent studies dealing with alternative habitats. Here it is difficult to separate that which is factual from that which is merely speculative. The feasibility of underwater structures as a new habitat for man is questionable, and it may be justified only in very unique circumstances.

For certain building types and even small communal groupings (such as airports and speedboat terminals, power plants, experimental stations and aquaculture buildings) floating or marine stable structures *do* offer significant advantages that may lead to their occasional implementation in future years. In fact, it is only through this type of shelter that any permanent "underwater" living with easy access to above water levels could become a viable alternative for permanent human habitation.

Other proposals that remove the human shelter from its natural earthscape include aerial structures: solar-powered aerostats (similar to the floating spheres proposed by Buckminster Fuller and based on the principle that a large enough body enclosing air would eventually float once the sun had heated the air contained within it) and other lighter-than-air structures, which may eventually be employed as special-purpose manned stations.[17] Also, more and more plans are being developed for (and around) outer space colonies. We will touch on these in more detail later on.

Consideration has been given to bio-buildings or buildings that grow (based on the genetic alteration of plant life to form shelters), crystal dwellings (based on the hope of eventually developing crystalline structures)[18] and also experimental ice buildings developed by spraying water mist on a framing metal mesh at freezing temperatures.[19]

There have also been attempts to develop three-dimensional light-solidifying plastic structures built through the projection of holographic images. The same technique has been proposed using sound vibrations in lieu of light rays.[20]

These technologies are all in the experimental stages and their success and feasibility are far from proven yet. In future years, however, many of these incipient methodologies, aided by computer-developed shelter design, may offer new alternatives to human habitats throughout the world.

COMMUNITIES, CITIES AND COMMUNICATIONS

There are few feasible alternatives for the expansion of small communities throughout rural America in future years other than horizontal or vertical growth. Few of the alternatives contemplated to accommodate urban expansionism would be feasible on the small community scale.

The future of small communities, however, may depend greatly on the specific region they are located in. It is very unlikely that the near future will produce agricultural, socioeconomic and industrial changes of the magnitude needed to revolutionize entire North American occupational and communal distributions on a regional basis. The increased urbanization, industrialization and economic growth of many sunbelt areas would not bring about changes drastic enough to obliterate communities and urban nuclei in the breadbasket region or to deindustrialize the Northeast and the upper Midwest. Hence, the most serious problems to be faced by most American towns and cities in the future will probably be the proliferation or worsening of issues already troubling large urban nuclei.

The possible patterns of growth for future American cities are: horizontal, vertical, a combination of vertical and horizontal, or underground. All of these possible expansion patterns, however, would continue to strain the already outgrown utilities and services of most American urban conglomerates. Thus, the only really effective measures that could be taken to accommodate any of these growth processes would be those also aimed at resolving the strain that edifices place on services, utilities and communication systems.

Based on the premise that the problems of overgrown urban nuclei are almost insoluble once certain population levels or densities are attained, many arguments have been made for the exploration of new frontiers to accommodate (or alleviate) urban growth, some even departing from the traditional land-based human habitation patterns and speculating about floating cities, underwater communities, airbound stations and outer space colonies. Other proposals have also included the decentralization and dispersal of communities (or even cities) to unpopulated areas (arctic and desert regions) or underground developments.

There is one point that must be made regarding all of these possibilities prior to analyzing the relative advantages and disadvantages of each: although long-range visionaries may conceive (and even justify) underwater, outer space, subterranean, or floating cities, urban conglomerates are never built overnight and rarely with a preconceived *total* space/function layout in mind. Throughout history such monumental tasks have been attempted seldom and unsuccessfully, such as in Brasilia and Chandigargh.

The birth and development of a community (and eventually of any urban conglomerate) is an organic process that is rarely generated by preconceived notions and intellectual assignments to alternative life-styles. Most important: all communities *must* have a purpose. The eventual development of some of these future "cities" could only occur through the growth of small communities that must begin with a clearly defined function. No one will ever move permanently underground or to outer space for long periods of time "just out of novelty" or merely "for the fun of it." Human nature is not comfortable with radical changes. Relocation in an underwater community because work on underwater mining activities and the life-style of an underwater community may be more appealing than a low-paying job or an over-crowded urban tenement could be a possibility in some cases. Unfortunately, competitive levels of amenities and appealing working and living conditions are never the case at the start. Many of these incipient communities would be likely to begin merely as experimental bases of temporary occupancy, and only in the event of their gradual success could the first features of actual community life be ultimately developed.

So, when referring to these new frontiers, it must be understood that many of the future scenarios described in the following pages may be extremely distant ones. The possibilities, however, *do* exist and must be considered.

It has been argued that the future of many American cities will depend on which collective attitude is adopted by society as a whole in future years. If the return to nature and other antitechnological movements dominate, it would be highly unlikely that anyone will see mega-projects sprawling throughout the urban scene. If technology dominates, then the result could be exactly the opposite.[21] Both of these proposals, however, represent simplistic absolutisms that are very unlikely to occur. The proper application of technoscientific advances *in conjunction* with collective environmental awareness would be the attitude most likely to dominate future years,[22] since it is the only logical alternative to the spatial and environmental adequacy of future built environments.

With regard to the patterns of development most likely to be followed by existing urban nuclei, several alternatives have been explored. The partial (or full) enclosure of many cities has been repeatedly proposed. In many cases, this would be highly convenient since it could reduce the energy consumption and climate/weather exposure of urban habitation units significantly and thus provide year round environmental control.

Year round environmental management could also be highly beneficial in the development of agricultural and industrial systems, and it could also provide for significant increases in habitable space within fixed spatial parameters. Independent urban entities (large entertainment, shopping or other special-purpose complexes) could also develop in this fashion. Presently, for example, shopping malls are already preparing the way for the proliferation of this type of sheltered urban complex.

The eventual enclosure of many urban nuclei may not necessarily occur in the form of overbearing geodesic domes spanning the urban landscape from one end to the other, but rather in a series of partial enclosures (domed, planned unit developments or buildings linked by aboveground or underground connectors). And, as urban sectors were enclosed, some transportation arteries would eventually be phased out. Since decaying areas would rarely be connected to the enclosed networks, their deterioration would accelerate, thus allowing for rapid renewal processes. Another advantage of this type of urban evolution is that it could be directed either inwards or outwards of existing urban parameters while still allowing for the traditional patterns of urban growth: horizontal, vertical and underground.

Projects dealing with the evolution of urban nuclei in America, however, must first be proven not only adequate but also feasible. If adequacy can only be attained through large projects aimed at modifying entire cities at once, it is very unlikely that any urban enclosure plan will succeed in the foreseeable future. The capital expenditures—no matter how "theoretically" feasible they may be—would be enormous for *any* nation in an overpopulated world with limited resources. But if their adequacy and feasibility are tested on smaller scales, then urban enclosed networks may very well offer a response to many present urban/suburban conflicts and provide a viable setting from which to welcome new horizons.

Long-range visions of future cities contemplate a myriad of alternatives, one of which is the development of aboveground mega-structures: large city/building complexes built over large piers and columns with extensive green belts both at ground level and throughout the main body of the complex of edifices.[23] Specific conceptions vary from the "arcologies" of Paolo Soleri to terraced and multilevel cities, and even some proposals based on biological growth concepts using combinations of natural and technoscientific processes.[24]

The concept of aboveground mega-structures is, of course, a very old one and will probably remain with mankind forever. The specific applications of many of these new approaches to urban conglomerates could redefine urban landscapes eventually. Here again, the logical patterns of growth (from small special-purpose nuclei to larger conglomerates) are likely to be the determining factor. A summary of aboveground mega-structure types includes large individual units linked together by massive transportation networks, combinations of aboveground (or above-water) and underground (or underwater) edifices, large domed or canopied units, or a combination of several of the previous types.

One variation on some of those concepts has been introduced with the proposal of mountain cities and cities built in abandoned open mine pits.[25] Presuming a relative independence from costly utilities and communication arteries, the obvious advantage to these proposals is their minimal environmental impact and the development of habitable spaces on lands unlikely to be used for food production or resources exploitation. This large scale version of the earth-sheltered dwelling could also become popular should the availability of land decrease sharply in the years ahead.

A more distant prospect is that of underground cities. Many present urban nuclei may already be incorporating many of the necessary features that could eventually permit a transition to this type of setting for human habitation. However, total underground living would be justifiable only in certain regions of the globe. The development of an underground city is a completely different issue from that of an isolated building. Problems concerning general emergencies, crime, services, ventilation systems, regulatory matters, property ownership, disease communication and a myriad of related concerns surface immediately when large concentrations of individuals with limited life support systems, spatial parameters and points of access are proposed.

The move to underground developments could probably involve transportation services and industrialized operations primarily in future years, and may eventually include many entertainment facilities, mechanical installations, power plants and even some habitation units. The evolution from these special-purpose entities to full community living, however, is another matter.

Marine communities and other oceanic special-purpose complexes are another possible alternative for the future. Japan has already launched a habitable structure above water and many of its urban expansion plans contemplate extending Tokyo into the Bay. Plans have also been developed for a marine city; and in England, a "Sea City" that could hold up to 30,000 people was recently proposed. Buckminster Fuller, a pioneer in this field, also conceived "terra city," a marine urban complex towering over 9,000 feet above the sea and capable of holding up to one million inhabitants.[26]

Aquatic habitats offer significant promise in the years ahead as the exploitation of vast ocean resources grows. But whether the development of these marine communities will become the responsibility of architects and urban planners or of nautical engineers is another matter. Already, many ships and submarines could fall into the category of oceanic "communities."

Aerial communities constitute an interesting chapter in the study of possible human habitats. Though their existence may appear very far-fetched right now, the eventual linkage of several lighter-than-air systems could conceivably result in the development of permanent aerial stations. However, once the planet's surface is abandoned, it is the space colony that surfaces as the ultimate frontier for human habitats.

Regarding outer space colonies, three possible sitings have so far been identified: low Earth orbit (250 miles up), geostatuary orbit (22,500 miles up), and near-Moon's orbit locations. Many of the problems imposed by Earth's atmosphere would not be present (rain and snow, pollutants, corrosion, vibration, earthquakes) but there would be some inherent limitations to the physical development of outer space structures: lift-off forces, intense sunlight and, above all, zero gravity. Although the structural properties of many members could be significantly increased (an aluminum beam, for example, could be miles long) and their weight would become insignificant, the assemblage of any structure could become a most tenuous and arduous task. Astronauts have been known to work extremely hard at performing the simplest chores precisely because of zero gravity conditions. Automation and preassemblages are the methodologies presently being pursued to resolve these problems.[27]

The possibility of outer space colonies is a very real one. Already there are plans for communication networks, planetary exploration stations and laboratory and manufacturing systems whose dependency on outer space colonies is fundamental (recent outer space repair and maintenance tasks executed by space shuttle crews are a clear example of these activities), and many studies have gone so far as to study the self-sufficiency and socioeconomic implications of outer space communities, and even to establish comparisons between possible volumetric designs.[28]

As the ultimate frontier, outer space habitation may, in fact, be a lot nearer than many people choose to believe. However, whether access to this type of community will become available to everyone, or whether life-styles and habitats will develop in a fashion even remotely similar to those in other Earth-bound special-purpose habitation settlements, is a completely different question.

Transportation is an issue whose potential development may revolutionize this nation's urban and rural landscapes. Many of the solutions to present urban problems as well as the adequate evolution of most American cities depend on the way in which transportation systems and communication channels evolve in future years.

On a small scale, moving sidewalks, public escalators, improved short distance mass transit shuttle systems, and other recent developments in public transportation hold significant promise. There also seem to be a multitude of alternatives to improve transportation on a regional and even on a national scale, as we will see later on. However, it is on the popular range of travel from one to two hundred miles (where the private automobile presently dominates the American transportation system) that more alternatives are needed.

In this respect, several proposals such as personal rapid transit, electrical monorail shuttle type systems run by computerized controls that stop at predetermined locations, and other similar systems have been proposed to serve as many as 29,000 people. For longer distances (especially in congested urban sectors) aircraft capable of vertical takeoff and landing, new dirigible aircraft, improved monorail and subway systems, and even magnetic trains capable of traveling at speeds up to 300 miles per hour are being considered and even introduced in many urban nuclei.

Nevertheless, not all mass transit systems implemented recently have succeeded. New transportation technologies are not easily introduced into the urban fabric or readily accepted by the average urbanite. In the area of physical travel, the two major improvements of the near future would lie in the development of electrically operated, nonpolluting personal transportation vehicles combined with a decrease in the travel patterns of the American population because of improved audiovisual/electronic communication systems.

If to these major improvements one adds the replacement of vehicular traffic in congested urban settings by intra-city shuttles, subways, waterways, moving sidewalks, monorails and other mass transportation systems, the future outlook for transit patterns in American cities begins to improve.

Such developments would resolve many transportation and environmental problems and would also permit more adequate usage of the excessively high percentage of valuable urban land presently allocated to inefficient transportation units. It would also be the only way many cities could eventually develop environmentally controlled enclosures or achieve underground expansion.

On a national scale, proposed transportation systems of the future include coast to coast underground travel in as little time as 54 minutes through the use of national subway networks linking major American cities. Subterranean cars, electromagnetically propelled and traveling in underground evacuated tubes following the Earth's curvature, could travel as fast as 14,000 miles per hour.[29]

Although still in the experimental stages, such environmental impact-free advances in transportation would revolutionize not only urban and suburban landscapes on a national scale but international travel patterns as well, since, in some cases, land travel would end up being faster than flying. Several tunnel corridors linking the Northeast to the Midwest, and the south central, southwestern and western states, have already been considered. Additional plans propose rapid transportation systems of this same nature that could also transport the passengers' private transportation vehicles.[30]

In future years, the "insoluble" problems that plague urban and suburban settings today may be overcome by developments presently regarded as only long-range visions. The solution to many modern calamities may actually lie in the move towards some of those new horizons.

Such a move may already be under way. Several of the systems and products needed to support many of these "futuristic" concepts are, in fact, being developed right now and in some cases are readily available.

16
Systems and Products

The American construction industry has increased its stock of systems and products significantly since the Industrial Revolution years, but at the same time it has also retained almost all of its traditional building procedures and materials.

Upon reviewing the sizable number of construction systems and products commercially available today combined with those presently in the experimental stages, one inevitably ends up asking the questions: why has the construction industry retained so many traditional characteristics despite all the systems improvements and new products that have been made available in recent years? Why has it not been radically transformed by the ongoing scientific revolution of the 20th century?

The fact is that it has been, but not as drastically as other areas of human achievement, because despite many applicable technological breakthroughs, building construction has been forced to retain the majority of its old ways and means. The reasons for this are complex, but in trying to explore them several issues *do* come to mind which may help shed some light on the matter:

1. The construction of buildings is an inherently "traditional" process that does not readily adopt drastic changes because most of its operations are

tied to the production of a durable product that, in itself, has remained relatively unchanged: the human shelter. Thus, when confronted with basic choices, building construction has often subordinated technological advances to traditional processes.

2. Because of the inherent durability qualities presumed and demanded of most buildings the construction industry is continuously forced to employ long lasting and "established" materials and systems, and generally tends to resist innovation. Most of the new products developed by man have met almost every qualification except that of durability, since for modern technology in general longevity is not an essential determinant of value.

3. Economics are also closely related to these issues. The long-term financial arrangements, feasibility and economics applied to the development of habitats are characteristically different from those used in other industries. Many of the products that will be mentioned in the following pages, for example, are being researched and developed for uses which would make them completely unfeasible in average building processes today. The experimental laboratories at the National Aeronautics and Space Administration (NASA) may be developing wonders that have the potential to revolutionize future building construction, but just how realistic that assumption is if there are no everyday feasibility or longevity tests, marketplace competition or even regulatory constraints applied to these promising materials and systems is an open question.

4. In many ways the fragmented structure of the construction industry may have forced it to resist excessive technological infiltration and thus slow down its relative rate of development. Innovations are always easier to implement when the decision-making process is in the hands of a few. The need to tailor its final product to everyday usage by all kinds of people and the safety considerations and regulatory issues that have increasingly intruded on American built environments are also significant factors.

Of perhaps less significance but still worth mentioning are the effects of regional and climatic characteristics upon edifices and the public attitude towards the development of shelters. With only a few exceptions (generally involving the outright unusual), "futuristic" building systems and products are only commonly employed in those building types that lend themselves to innovative environments, such as modern office complexes and specialized factories and communication buildings. However, most building types are not "future" oriented at all. Also, because of the inherent longevity of human shelters, many edifices (still perfectly adequate for modern living) were built at times when innovative systems or products were simply nonexistent. And because of the constant need for improvements and adjustments necessary to incorporate new features in preexisting built environments, do-it-your-self processes, which cannot easily incorporate highly special-ized/advanced operations and thus must rely on simple procedures, have further facilitated the retention of many traditional building systems.

SYSTEMS

The most significant innovation of the next few years in the area of new systems applied to the development of edifices will probably be the increased use of computers as an aid in the design and production of shelters. There are, of course, inherent advantages in computerized building designs and automated construction: improved ergonomics and efficiency of spatial design, speed of execution and better control of building processes.

Electronics and automation may also open the door to new structural concepts and building systems presently considered merely experimen-tal: cellular developing structures—spaces whose structural supports also act as room dividers—which can grow into large complexes whether by accretion or by clip-on or slide-in procedures around central stable cores; space and marine structures; long-span or underground architec-ture; lighter-than-air dwellings; biostructures; or even shelters devel-oped by chemical processes (crystallization, for example).

Automation and electronic construction systems could also facilitate the popularization of many emerging building systems. Overhead membranes, for example, which only recently began to appear in ur-ban landscapes, have proliferated remarkably within the past few years due to advances in the physical properties of the materials employed for their construction, and also because of a much better understanding of cabled structural designs. Fiberglass yarns coated with Teflon resins (virtually nonflammable) over steel pylons have been employed to cover 150 acres (an area equivalent to 80 football fields) over the new Jed-dah International Airport near Mecca—an enclosure that would roof 25 percent more space than the largest building in the world, the Pen-tagon. Other structures of this type built recently include large mem-brane roofs over football stadiums and big department store com-plexes; and presently, more and more projects are being proposed to cover entire streets in urban sectors.[1]

The key determinants for the successful execution of building projects lie basically in *what* to use (materials), *how* to use them (system) and *when* and *where* to do so. Changes in materials availability, or in requirements relative to their durability and workability, as well as new tools for working them, changes in transportation systems and erection procedures or the new design possibilities based on those changes could conceivably transform the construction industry in future years.

Full scale prefabrication procedures, the development of lighter structural systems, mobile and fully automated factories or assemblage techniques, robotics and air-borne transportation may seem to be far in the future, but in fact their implementation could be very near.

Presently, one of the most complex problems to be resolved in the use of prefabrication components is the size limitation imposed by ground transportation restrictions; no single unit can exceed 12 feet in width or 40 feet in length if it is going to be transported on any interstate highway. Helicopter lifts, dirigibles or helium-filled aircraft could easily circumvent these physical restrictions.[2] Experimental spherical, helium-filled aircraft units recently developed in Canada, for example, have demonstrated that uplifts of up to 40 tons could be achieved.[3]

Other newly proposed systems and materials could eliminate prefabrication altogether or use prefabrication processes to simplify assembling buildings on site as do some recently developed techniques which form dome-like structures with simple structural members covered with a fiberglass mesh and sprayed with insulating foam.[4] Spraying as an integral construction system has also been used for the construction of temporary and experimental ice buildings by spraying water mist over steel mesh frames at subzero temperatures.[5]

There are other assemblage techniques on the horizon which may eventually change the traditional prefabrication concept or even on-site construction procedures by blending technologies from each fabrication process within new building techniques. Filament wound systems, for example, involve the winding of strands of resin-coated glass filaments around inflatable or collapsible forms. Similarly, a construction system developed by the Midwest Applied Science Corporation (MASC) is capable of actually extruding buildings by spinning out plastic forms that rise and harden rapidly.[6]

Many of these systems, however, have been around for well over a decade without having a significant impact on building construction yet.

Since the early 1970's, building techniques have been proposed involving truck-mounted movable booms capable of extruding foam plastic that hardens in less than 10 seconds and is guided into place and physically "molded" by steel plates to form walls and ceilings. And the development of shelters through the use of inflatable and/or collapsible forms has been around for a couple of decades.[7]

Nevertheless, it is undeniable that, when added to the increasing use of other construction systems (modular units, prefabricated components, mobile bridges, and improved transportation and assemblage systems), many recent developments could change the physical development of shelters throughout the industrialized world in future years.

Fiber optics and holography linked to materials that react to light, laser beams and rock-melting excavation technologies, improvements in concrete admixtures and entraining processes, wood treatments—all of these could also help systems and construction techniques. Only recently, for example, remarkable results are being achieved by the use of fiber-reinforced concrete—steel, glass or plastic fibers added to concrete mixtures—that reduces the amount of concrete needed by 50 percent while providing better flexural strength.[8]

On a collective scale, solar technologies, nuclear fission, Earth-orbiting solar power generating stations, ocean thermal energy conversion plants, and continental energy supply networks could solve the energy demands of American habitats. Recently, an Arizona building firm proposed an all-solar housing development with electric cars for local transportation as built-in appliances. The homes would generate more electricity than they would actually need, so they could feed the excess into a local utility grid that could charge the electric car batteries.[9]

Similarly, the development of water and sewage disposal systems could be worth watching in future decades. Water efficient plumbing could eliminate much of the water waste presently affecting most housing units. An average water closet, for example, uses up to 6 gallons of water per flush cycle, while an efficient one (still employing a water-flush system) could use as little as 1.5 or 1.0 gallons. Many average units currently available already have improved water consumption demands, using anywhere from 3 to 0.1 gallons per flush cycle (as in the case of detergent-flushed units), and presently even some waterless toilets are beginning to be marketed. Water-saving faucets may also help decrease the number of gallons consumed per minute from 2.0 to 0.75

through reduced-flow inserts; and the same principle applies to shower heads, where water consumption drops from 3.0 gallons per minute to 0.5 gallons per minute when air-assisted low-flow technologies are used.[10]

Further improvements in these systems could lead to in-house sewage treatment and recycling processes, or to the preconditioning of gray waters or raw sewage prior to their disposal in public or private treatment systems.

It has been argued that advances in building systems and the further simplification of many construction techniques may also facilitate the practice of do-it-yourself housing. However, this proposal has some negative aspects worth mentioning.

No matter how simple some of the building processes may seem to a layman, it *does* take a certain amount of training, experience and skill to carry them out. A recent study by the United States National Bureau of Standards, for example, revealed that 87 percent of the fires caused by wood stoves were the result of improper installation, maintenance or use of the units.[11] The high cost of traditional heating sources and the apparent simplicity of installation and operation of wood stoves induced a multitude of unqualified users (up 40 percent in 1979 from 1978 alone)[12] to install these items in their homes, most of them following do-it-yourself techniques that were neither properly understood nor adequately executed and most of whose end results were neither inspected nor certified by qualified contractors.

Moreover, the legal implications of do-it-yourself construction go far beyond its immediate use since liability can easily be extended to cover long-range deficiencies or malfunctions, and there is neither adequate legal protection nor established guarantees or precedents for guidance in such situations.

The fact is that in building construction reality rarely follows long-range forecasts. Many functional innovations in certain building types, for example, have encountered resistance in the interwoven legal aspects of preestablished systems and traditional operational processes despite their obvious advantages. Almost two decades ago, "controlled environment capsules" were proposed as a viable possibility for solving a multitude of operational problems in health care centers. Automated fireproof transparent capsules (which could be polarized to insure privacy) moving within internal overhead rail systems would physically transport patients from one area to another, thus solving the isolation, communication and adequate monitoring problems of hospitals through controlled environments, ease of mobility and immediate audiovisual communication with all patients.[13] Over 15 years later, however, not a single hospital in this nation has adopted a system that, even remotely, resembles these operational innovations.

Similarly, many breakthroughs in construction systems have not met with an adequate response in the building industry, despite their apparent advantages. Buckminster Fuller's dymaxion house, a factory-built, lightweight metal dwelling constructed around a central utility pole that could be literally flown to job sites, was developed well over half a century ago, and his geodesic domes—half spheres resulting from the joining of several tetrahedrons—a structural design shape that encloses more space per unit of weight than any other known to man, have been around since the end of World War II. Neither system, however, has had any significant impact on everyday construction.

The assemblage of these structures is surprisingly simple: the triangular framing members of geodesic domes, for example, can be made out of wood, steel or aluminum pipes and even of bamboo; over these, any variety of materials can be laid to form an outer skin (galvanized steel, aluminum, wood, Teflon-coated membranes, plastics, or even treated cardboard laminates). Many of the structures conceived in this fashion have the added advantage of not requiring highly skilled labor for their assemblage, so that they could be made available in simple do-it-yourself "kit" forms and would require almost no expert help, except for possible inspection or certification of the final product.[14] Despite all of this, however, the use of these systems has not altered most traditional building processes, nor has the use of geodesic domes revolutionized the developments of this nation's built environments.

Unquestionably, there is an inherent relationship between new building systems and the products (or substitutes) they employ, and despite the dependence on traditional materials, the growing variety of products serving the American construction industry seems to be following trends which hold significant promise for future years.

MATERIALS AND PRODUCTS

The selection of materials and products for any building project is determined by four basic factors:

1. Performance. The materials or products will have to perform satisfactorily under all the conditions to which they will be exposed.
2. Adaptability. They must lend themselves to the changes and adjustments needed to make them work within an organized system.
3. Life expectancy. They must be reliable enough to perform a specific function or maintain their physical characteristics under certain conditions for a defined period of time.
4. Feasibility. They must be adequately priced relative to the other three factors.

Many of the significant developments relative to future habitats, therefore, will depend on the conformance of new building materials and products with those essential factors. Change, however, may redefine the applications of many value concepts or introduce new ones. Until a few years ago, speed of execution was not as essential in construction as it is today, and had little to do with the feasibility of a project. Presently, however, speed is essential in building construction.

In years to come, it would not be surprising to find computer terminals at job sites, and the processing and recording of building documents and trades communication done entirely in this fashion. Tool improvements—until recently a relatively stagnant area in building construction—are already beginning to expand from the areas of "specialized" artifacts developed for specific building systems and are starting to change basic procedures, such as sawing, nailing and other mechanical attachment systems. Just recently, for example, it was discovered that by merely bending the handle of a conventional claw hammer slightly, the pounding effect of the tool on the wrist could be significantly reduced.[15]

Several new construction products are also being developed. A new sulphur-asphalt mix has been proposed as a paving material that would not soften in summer or crack in winter, would help prevent new shortages of petroleum-based asphaltic mixtures, and would have the added advantage of providing an adequate use for the millions of tons of excess sulphur burdening this nation every year.[16] New lightweight alloys and "metal-matrix composites" (metals inside other metals) are also producing combinations of significantly improved weight/strength capabilities with added plasticity and weather resistance.[17]

Plastics, in general, are among the most promising construction materials. Already components stronger and lighter than steel have been molded to suit many building features, and many finishes and cementitious coatings have also benefited significantly from the introduction of plastic additives.

Wood is a traditional construction material that is being continuously improved through better treatment processes; its availability also could be significantly increased by the introduction of new tree harvesting techniques. Recent experiments financed by the Department of Energy have yielded promising results in this area by using existing root systems and stumps of harvested trees to grow new ones instead of replanting with seedlings, thus reducing the growth time by as much as 50 percent.[18]

Photovoltaics and other solar technologies are also slowly redefining many traditional building features. Already available or in the process of development are solar roof shingles, improved building thermal skins, and computer-controlled thermal membranes that maintain or alter constant indoor environments, no matter what the temperature is outside. Also recently developed are polystyrene building blocks which meet all state and federal fire resistance and strength requirements and produce super-insulated edifices with virtually no need for heating or cooling of any kind. Heat is provided by the occupants themselves and by electric appliances and lighting, and is easily retained by the shelter, while cooling is achieved by carefully developed ventilation shafts. Significant advances have also been achieved recently in temperature control through improved glazing elements, sun operated/sensitive glass, heating/defogging glazing units, etc.

New paper derivatives and other synthetic materials in the form of honeycombed or laminated panels are already in common use in many modular construction systems, temporary structures and interior partitions. One typical example is a recently introduced panel made out of resin-stiffened paper sandwiched between two fiberglass layers which make it completely water resistant; pound for pound this panel is stronger and cheaper than steel.[19]

Improved noise control could be achieved by employing new lamination systems; thin sheets of steel and plastic laminated together have already been developed by the steel industry as an excellent noise insulation component.[20] This could prove to be an essential necessity if environmental noise continues on the rise (it has increased 100 percent every 10 years during this century).[21]

Paint and other special coatings have also recently broadened their scope in the building industry well beyond the provision of ''color'' and expanded into specialized areas including textures, weather/corrosion resistance, thermal/electrical conductivity, psychological and therapeutic effects, pest control, health and hygiene and many other areas. In future years it would not be surprising to find that special coatings had redefined buildings' outer skins. Already, special-purpose coatings are essential elements in electrical installations, radiation and insulation protection, fireproofing and anti-graffiti finishes. In fact, with the recent development of nonfading colors and integral color processes, many new and improved special coatings could revolutionize many building features in the next millennium.

Electronics are also influencing various building components. Already commercially available are speech synthesizers ready to be installed on any elevator or vertical transportation module. Developments of this nature could result in the introduction of actual voice-operated systems in buildings throughout the nation with special features that could include taped messages and directions or even instructions in case of emergency. Similar technologies are being explored for general exiting and fire emergency procedures in new buildings.[22]

Electronic innovations are also apparent in new household appliances and equipment that save space and time and that are progressively becoming more automated. Although robotics may eventually redefine many housing features, it is also possible that *totally automated* shelters will be developed, thus permitting many household tasks to be done by the shelter itself. That is, man would not only be served by robots but actually *live* within them.

Utilitarianism must not be totally disregarded. Especially interesting is the proliferation of furnishings and household items that serve dual/triple functions, save space, or merely convert into completely different items once their primary function has been served.

Solar technology and other recent technoscientific developments are also apparently influencing the appliance and household equipment world. Already proposed are solar stoves and furnaces, chemical lighting systems and even nuclear powered refrigerators,[23] as well as solar powered roof vents and exhaust systems.

However, not all new products that originally appeal to forecasters dealing with future edifices result in widespread or revolutionary implementations. About 10 years ago, ''structural foam buildings''—interlocking foam units with cavities containing reinforcing steel rods every 24 inches, eventually pumped full of concrete—were proposed as a promising construction system that could reduce the execution time of housing units to approximately 300 man hours.[24] Their use, however, never became widespread.

One of the problems is that the future implementations of many new products are not as straightforward as initial forecasts tend to presume. Over 10 years ago, "fibershell," a new hybrid material consisting of a resin-impregnated, cellulose-based honeycomb core and woven fiberglass facing bonded to gypsum wallboard panels for interior uses, and to treated woods and other weather-resistant materials for exterior facings, was proposed.[25] Despite its significant advantages over many existing panel wall systems, fibershell has not become a common building material.

Nevertheless, there are some products presently in experimental stages that hold significant promise. Among them are several innovative building materials such as those derived from electrolysis processes capable of actually "growing" underwater structures through the adherence of mineral deposits on electrified metal frames that produce concrete-like substances 30 percent stronger than average concrete mixtures and capable of developing certain structures more inexpensively than with conventional technologies.[26] Projects that range from the development of specific building components for use both above and under water, to entire structures or even islands are now under consideration.

Other technologies include the use of sprayed plastic molecules that solidify once in contact with strong light, in combination with computer-operated building machines that control the sketching, design and projection of three-dimensional holographic images, thus transforming mere conceptualizations into physical structures of whatever scale or size desired in a fraction of the time presently dedicated to shelter construction.[27] Similar propositions have also been considered for remodeling existing structures and for underwater or even outer space building technologies.

Among the basic reasons why many new products and systems are not readily accepted by the American building industry are the questions of limited feasibilities and applications and the conflicts arising from the inherent conservativism of an industry that requires high investments. An additional problem is the sometimes limited availability of the specialized labor required by many new building processes.

There are also the enormous burdens imposed by regulatory issues. Building codes throughout the United States classify building types according to size and function. Each structural component of these building types must then conform with specific requirements for fire resistance, exiting, overall height, number of stories, etc., depending upon their classification, location within urban "fire zones," zoning restrictions and a myriad of related factors. To comply with these interwoven restrictions, many new products end up being combined with systems and materials previously "certified," thus sacrificing much of their effective feasibility to the regulatory mandate. Because of this, many are set aside as unfeasible technologies or just plain "luxury items."

This should not be taken to mean, however, that all new products or processes proposed as likely building materials or systems follow this course, nor that their ultimate success depends primarily upon their triumph over a complicated regulatory matrix. It is noteworthy that many new and "revolutionary" implementations in building construction have not lived up to the expectations placed upon them by their developers or their users. Most recently, an all-plastic foam modernistic house built in 1975 was put up for sale as "in need of repair due to vandalism." [28] Obviously, the test that this shelter failed to endure had little to do with a building code.

Barring excessive increases in regulatory constraints and taking into consideration the inherently slow evolutionary process followed by building construction, an overall analysis of the future trends of the building systems and procedures presently employed in the United States shows that the following components or processes are likely to gain significance in future decades: heating, ventilating, air conditioning and other artificial climate technologies; electrical/electronic installations; telephone/audiovisual and other communication systems; plastics and synthetic fibers; chemical fasteners, glues and adhesives; special-purpose coatings and paints; prefabrication and preformed assemblages; conveying systems; aluminum products and installations; special construction and equipment; wood and wood treatment processes; new glazing techniques; insulation and noise control products; and paper or paper derivatives applications.

Conversely, among those whose influences in building construction could easily decline in future years one could mention: traditional plumbing systems, masonry, cast-in-place concrete, traditional roofing and paving systems, rough carpentry, copper and galvanized piping and on-site steel placement.

Materials shortages, however, may drastically alter the relative evolution of many of these individual processes, just as new recycling procedures and the need for the remodeling and preservation of existing structures will maintain the need for many traditional building processes.

Unforeseen events, shortages or crises may also introduce or popularize new technologies. Just a few years ago a 700-square-foot living structure was developed in Troy, New York, built out of cardboard tubes, beer bottles and aluminum cans set in mortar, with the roof made out of discarded rubber. All the materials were obtained at a cost under $700.[29] And researchers at Purdue University have recently developed brick units using sewage sludge: "bio bricks" that have the same strength, appearance *and smell* of other bricks.[30] Recycling processes or the emergence of new building technologies derived from the introduction of discarded items as possible construction materials cannot be completely ignored.

To erect his first shelter man had to develop a building system. In doing so, he selected materials and developed tools with which to accomplish the task. But the conceptual shelter did not originate there. The spatial and environmental adequacy of habitats as well as all the tools, products and building systems known to date have also been defined by socioeconomic, political and cultural features.

The ways in which these interactive factors could organize themselves to stimulate or unnerve many of the issues and characteristics previously mentioned is the subject of the next chapter.

17
Organizations

The different ways in which future societies are organized will provide the blueprints for future communities within existing habitable environments or in foreign mediums.

Changes and evolutions in the organizational patterns of most societies, however, are not easy to prognosticate. Sometimes even simple short-range forecasts of relatively minor changes in life-styles turn out to be extremely inaccurate assessments. In April 1978, a hopeful article illustrated the likelihood of increased work-at-home patterns and use of home computers in a scenario that involved a secretary doing almost all her work through a remote dictation unit and a computer terminal installed in her home, speculating that this could become a common situation in 1983.[1] However, life-style changes of this type are still rare and have certainly not become common for clerical employees. As a matter of fact, today the likelihood that work-at-home patterns will become widespread is being seriously questioned, and with good reason. The loss of human contact, fewer links to society and reduced social life have been clearly identified as potential problems for work-at-home employees. Studies conducted at IBM offices, for example, have shown that contact between office workers decreases proportionately as the distance between them is increased and that floor to floor separations can cut contact by as much as 90 percent.[2] Other problems such as external interruptions common in the home, building and zoning restrictions and security factors have also been mentioned. Thus it has been generally concluded that work-at-home practices may not be as easily achieved by regular employees as they would be by independent consultants or subcontractors to large corporations, or by self-employed people.

One of the most significant effects of work-at-home activities could be the loss of household intimacy and home life. The absence of tangible breaks between business and home affairs and the loss of privacy to employers, fellow workers or even clientele could end up redefining entirely the concept of shelter and eventually change the traditional understanding of home from the place to retreat to after a day's work to a place where work, rest and entertainment are no longer functions of schedules and spatial settings, but of time lapses alone.

Forecasts about social developments are likely to influence the planning and development of pioneer communities in foreign mediums (water, air, underground, outer space) or even of new communities developed in more familiar landscapes (uninhabited or desert areas, decaying urban nuclei).

Planning decisions always involve risks, but at no time have the variables been so complex or the consequences so far reaching as they are today. Today more than ever before, man is forced to plan his endeavors on a long-range basis which, paradoxically, is the most difficult forecasting range. Future colonies and communities are no exception, despite the fact that settlements do not develop according to easily foreseeable evolutionary patterns or within the boundaries of predefined spatial parameters. In fact, in those few instances when evolutionary preconceptions or models have been imposed on habitats, they have failed.

In this respect, it is pertinent to point out that in habitats even many assumed standards are consistently undergoing significant revisions and adjustments. Until recently, for example, the standards set for lavatories were assumed to be correct; however, detailed analyses have shown that, based on their function, the average 32 inch mounting height for a basin is, in fact, too low and should be raised to approximately hip level (40 inches), while the bowl should be deeper, the fixtures recessed and the water-source level raised.[3]

One of the basic problems is that despite some general classifications that identify the main characteristics of most sociopolitical structures and/or communal organizations, the interactive issues that ultimately define a community are so overwhelmingly complex that their comprehensive description is almost impossible. In modern societies, two basic types of social systems have been defined that appear to include most organizational patterns: the unrestraining/structured type and the retaining/integrated type.[4]

The basic differences between the two are in the degree of mobility, homogeneity and flexibility of the social structure. It is relatively easy to move in and out of an unrestraining/structured social system because the association among individuals is *voluntary* and thus contributes to increased homogeneity. In the second social system type (retaining/integrated), however, there is a tendency to keep nonconformist elements within the system's structure and thus to promote social heterogeneity either by collective concordance or by totalitarianism (depending on the political philosophy).

Most American communities traditionally belonged to the unrestrained/structured type. However, from the beginning of the 20th century onward certain characteristics corresponding to the second type can be found in many urban nuclei throughout this country.

It is very unlikely that a nation born of an unrestraining/structured type of collective behavior, such as the United States, will produce organizations of a completely different kind. However, this type of social structure could pose some problems in the future if "planned" communities become increasingly necessary, because the preselection of social elements or even the careful planning of the evolutionary patterns of relatively homogeneous groupings could not necessarily guarantee social stability, nor would either of these alternatives accurately reflect this country's cultural fabric. Today more than ever before one finds increased interaction among sociologists, psychologists and designers in the development of new habitats. This trend may improve the adequacy and effectiveness of habitats, but it may also increase homogeneity through the introduction of sociological standards and behavioral averages in the conceptualization of built environments.

The possible types of future communal organizations, however, are not merely defined by cultural characteristics or political structures alone, but include a myriad of related subcategories.

Table 7
Communal Organizations Types

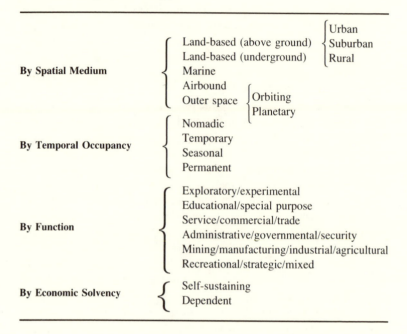

By Spatial Medium	Land-based (above ground)	Urban / Suburban / Rural
	Land-based (underground)	
	Marine	
	Airbound	
	Outer space	Orbiting / Planetary
By Temporal Occupancy	Nomadic	
	Temporary	
	Seasonal	
	Permanent	
By Function	Exploratory/experimental	
	Educational/special purpose	
	Service/commercial/trade	
	Administrative/governmental/security	
	Mining/manufacturing/industrial/agricultural	
	Recreational/strategic/mixed	
By Economic Solvency	Self-sustaining	
	Dependent	

Some of the possible classifications that could be considered according to different organizational features are shown in Table 7.

In each category, however, there is a vast difference between the planned communities and the realities of man living in new frontiers and foreign mediums. Ultimately, the success or failure of tomorrow's living environments will depend not only upon adequate sociological models or technoscientific advances but on the correspondence between human beings and their *total* living environments.

TRADITIONAL SETTINGS

Most large cities in the world are located in areas with a relatively moderate climate. Very few urban settings are located north of the 60th or south of the 45th parallels.[5] Thus, for all nations, the presence of inhabitable climatic conditions for at least parts of the year appears to be an important factor in the successful establishment of sizable communal groupings. Topography and accessibility also seem to play key roles in this respect.

The social, economic and political characteristics of groupings have also influenced the nature of urban conglomerates, and today one can define a city as a conglomerate of people in a specific location who can permanently govern and support themselves by their joint socio-political and economic activities.[6] The larger the economic base, the larger the community and the more complex its organizational structure.

Ultimately, communities are not rigidly structured entities but dynamic organizations containing a multitude of dynamic suborganizations, each of which is continuously undergoing changes and adjustments. Physically, a progression can be easily drawn from the single family dwelling to the neighborhood to the suburban sector and to the urban nucleus.

In these complex organizational systems even minor life-style changes may affect urban space distributions. The proliferation of swimming pools, for example, has consistently reduced the overcrowding of public beaches in states like Florida and California. During the last few decades Miami Beach has actually *lost* public beachfront without any excessive overcrowding on its public beaches, despite Florida's explosive population growth.

Life-style changes extend beyond the confines of household and community and eventually reach entire urban nuclei. Thus as broader levels are reached, new segments of the social system are affected. Shared housing units, for example, as well as multi-adult households or unisex families could have tremendous effects on communal settings in future years. Already in Australia it has been reported that, through a rental agreement with the state housing authority, many elderly persons live in relocatable garden cottages placed in the backyards of their children, grandchildren or other relatives. The prefabricated units are trucked to the site and hooked up to the plumbing and electrical services of the main house. They take approximately one day to set up or relocate.[7] Arrangements of this nature could become popular for these and other household types throughout this country in future years.

As electronics begin to invade household units and alter life-styles, new alternatives for work scheduling, shopping, banking, schooling, and even socializing could further reorganize many of the traditional space/activity layouts or even the construction characteristics of most American built environments. If the widespread usage of home computers increases work-at-home or home business type activities, for example, new fire safety requirements for some portions of future American shelters could be introduced. A set of "floppy" disks containing the essential core of a home-run business or work assignment, if lost in a fire, could cost much more than the mere loss of replaceable valuables; it could conceivably wipe out entire livelihoods.

Some recent forecasts of increased community-shared housing efforts and neighborhood/community cooperation in activities such as the provision of services, light manufacturing and even home building are highly questionable, because improved automation patterns have traditionally tended to decentralize physical contact among individuals and households alike, thus it would be unlikely for this nation to experience increases in communal activities as the result of an electronic "revolution." Increased self-help housing and remodeling due to improvements in building technologies are much more likely developments.

The concept of shelter portability, if further developed, may also facilitate the proliferation of movable communities; presently, except for certain agricultural groupings and some art/craft groups, this practice is almost nonexistent in America. Relocatable communal groupings could provide agricultural, entertainment, research, special services and even educational activities nationally on a cyclical basis in future years, and even serve communities developed in foreign mediums.

On the opposite end of the work-at-home spectrum, the general operational characteristics (scheduling, functional placement, etc.) of future organizations could also cause situations in which the traditional concept of shelter habitation would be drastically changed through the breakdown of sleeping, eating, working, studying and gathering activities in temporarily occupied areas intended to serve workweek-long periods only. This trend, rooted in the occupational patterns of the Industrial Revolution, has not ceased to gain momentum throughout this century. Urban apartments used only for sleeping purposes during the workweek by many modern dwellers who own "weekend cottages" in removed settings illustrate the strength of this trend.

The sociological and economic evolutions of the major American cities will have a significant impact on this nation's total built environment in future years. In this respect, the decline or rebirth of large American urban conglomerates are worthy of careful analysis. Among the many items cited as factors contributing to the future decline of American cities are: economic crises, suburban sprawl (both physical and economic), overcrowding, limited growth and building development, racial conflicts, crime, transit inadequacies, and deficient utilities and services. Future social, economic or occupational developments and changing life-styles may contribute to accelerate or retard the effects of present trends in some of these areas. In this respect, the potential influence of the mega-owner concept is a possibility worth considering. It is likely that large corporations, which are the only entities powerful enough to deal with complex regulatory constraints and large long-term investments or financial commitments, may end up owning large sectors of urban/suburban settings throughout America (or even on a worldwide basis linked to an international housing authority and resources management organizations).

This development, in turn, would encourage the development of mega-builder multidisciplinary organizations capable of handling all the real estate transactions, functions, designing, building and management of human shelters in future years.

Anything that would contribute to the unification of the elements presently involved in the development of this nation's built environments must be looked upon as a positive factor. In this light, some of the transformations previously described could resolve problems that have haunted the construction industry in America for decades. For example, a recent study has shown that scattered construction marketplaces and the high-risk and labor-intensive characteristics of the construction industry have seriously damaged the smooth flow of building construction in this nation for many years, causing a high degree of vertical and horizontal fragmentation as well as an almost chaotic discontinuity of leadership.[8] Thus, any movement towards unifying these functions under a multidisciplinary umbrella should be looked upon with interest, since it could provide a badly needed alternative to the otherwise seriously fragmented organizational structure of the American construction industry.

In many ways, the immoderate fragmentation of the elements involved in the development of habitats in this country is largely responsible for the excessive government interference in American built environments. Further unifications would probably help mitigate some of the regulatory restrictions presently placed upon this country's edifices. It is highly unlikely, for example, that regulatory provisions of questionable merit such as the "inclusionary" housing regulations—a system that requires a certain percentage of housing units for less affluent households to be made available at a lower cost than others in new developments, thus increasing the overall cost of the remaining units—could ever be forced upon new housing developments were the development companies large enough to dispute these impositions at the proper levels.

The problems posed by changing social organizational structures will be among the essential determinants of America's future built environments; but no matter how great their influence may be on the evolution of conventional habitats, its greatest impact will be felt on the communities developed in mankind's future habitation frontiers.

FRONTIERS

In the analyses of traditional communal settings, the impact of geographical factors on human built environments and habitation patterns is generally the starting point. Such is not the case in the studies of new frontiers presently opening to modern societies. In most of the new mediums to be explored by man as possible milieus for his habitats, traditional factors such as climatic conditions, topography, views and landscaping would be inherently subordinated to or circumvented by technoscientific advances, thus making the shelters' absolute dependency on technological implications their first unique characteristic.

But technoscientific principles and dependencies are not the only essential factors that may distinguish these entities from the traditionally known communities of man. One must consider that in these future communities spatial settings as well as environmental surroundings would be the result of value judgments, plans and decisions devised strictly by the human intellect. Thus the physical layout of the first communities established in most of these new frontiers will probably be based entirely on the presumption of how things "should be," with little room for error or readjustment. This issue should not be disregarded, since inadequacies affecting these pioneer communities could eventually determine the success or failure of new habitation alternatives.

Relative to the development of communities in new frontiers, it is worthwhile mentioning three basic prototypes of social organization systems (integrated, fragmented and aggregated)[9] and how each one could fit different kinds of communal arrangements.

Integrated social organizations strive for collective achievement through the generalized acceptance of objectives and behavioral patterns. Differences and individualistic traits are discouraged, and operational functions, spatial parameters and formal characteristics are developed, grouped and maintained in accordance with predefined and collectively accepted prototypes and standards.

Fragmented organizations concentrate on the individualism, self-sufficiency and personal independence of their elements. Predefined, patterned and rigid structures—whether social, political or economic—are undesirable, and achievement of individual goals is the main communal thrust.

Aggregated organizations are, in fact, a balanced mixture of the other two. In this social system, the definition of objectives is not normally achieved by accepting behavioral models only, but also by allowing individual characteristics to become part of a heterogeneous working entity, thus trying to achieve unity and progress through individual diversity. From the present standpoint, this last type would be the most desirable for the organizational structures of future communities in the United States.

Despite their extreme importance, however, social organizational features are not the only factors involved in the conceptualization of built environments. The environmental and physical characteristics of underground housing or outer space stations, for example, pose questions that go far beyond mere sociological speculations precisely because of the specific mediums and conditions under which they must develop.

The first issue to consider regarding future communities in new frontiers is the correspondence between their functional structure and their medium. Underground and floating colonies appear to be the ones most likely to develop in the near future because of the many similarities they have to existing land-based communities.

The full-time occupancy of underground communities, however, presents some serious obstacles which cannot be ignored. Many traditional characteristics of shelters and man/space relationships would have to be either altered or merely simulated; the lack of windows, for example (considering that windows are essential features of most dwellings that have gained importance over the years), as well as the redefinition of many other basic concepts dealing with "indoor/outdoor" relationships, could become serious stumbling blocks for adequate underground living environments. For one thing, the "outside" of the shelter would no longer lead to the open air and, in fact, upon leaving one's shelter one would be stepping from the *inside* into the *inside* (similar conditions and effects could also be experienced in "domed" cities). Thus the relative position of human beings with respect to their dwellings would be measured in levels of enclosure rather than in the simple in/out terms that presently define most architectures. Similar circumstances could also be encountered in future mega-building complexes or in "domed" urban sectors, although in those cases, the simple fact that the sky and outdoor spaces could be within visual reach might lessen the effect of these conditions.

Some of these potential problems may seem simple to circumvent at first, but the possible effects of these changed spatial conditions on human beings do pose some very serious questions relative to the ultimate adequacy of some of these habitation environments. Permanently occupied underwater communities, isolated from above-water or floating structures, are unlikely types of future habitats. The major problem of permanent underwater living environments is a question of logistics: besides the serious obstacles posed by watertightness, accesses and egresses, lighting the ocean floor at great depths in areas large enough to eliminate a sense of confinement would be extremely difficult and ecologically unsound, and in shallow waters submerged communities would be relatively useless. Except for very special functions, it is questionable that full underwater habitats will be developed unless they are linked to above-water (fixed or floating) surfaces and structures.

Outer space colonies are a separate issue and pose their own unique problems. One of the characteristics that immediately surfaces as a key difference between this type of habitat and others is the time factor involved in their occupancy. Even in the unlikely event that dwellers were submerged at great oceanic depths, access to the Earth's surface—with all its natural environmental characteristics—could be gained within minutes. In outer space, however, short-term occupancies (as presently understood) are an impossibility. Because of this, the development of outer space stations and communities will be dependent upon a multitude of time/habitation factors and special living circumstances not generally considered in Earth-based human-built environments.

Studies have been developed relative to the type of communal organization most adequate for these groupings (the type of social organization previously described here as "aggregated organization" has often been recommended), describing their optimum populations and population rotation depending on their specific functions (from 200 to 10,000 people for the initial ones),[10] their economic structure, and their spatial and environmental characteristics. In this last instance, it has been suggested that the smallness of the living quarters, the evenness of the environmental conditions and the predefined and unchanging boundaries within which everyday life would take place, could eventually affect most inhabitants significantly, making the continuous turnover of dwellers (even from community to community) a desirable feature.[11] This suggestion, however, which tries to circumvent the problem of physical confinement by giving outer space colonies the characteristics of temporary habitats and making nomads out of space colonists, could only be considered valid in very special cases, mostly defined by the colony's proximity to Earth.

Traditional forms, such as cylinders, spheres and circular torus shapes (the last of which are said to be most adequate) have been analyzed. Also, serious considerations have been given to the type of spatial distribution that should be followed in outer space living environments. In this area the analyses have covered the functional categorization of spaces and the development of utilitarian and noncategorical layouts, and have finally recommended combinations which may permit changes and adaptations from one scheme to another should the need arise.[12]

The actual construction of a space colony has been estimated to require—with available technological systems—approximately 2,200 workmen, and at present its proposed construction materials consist of an exterior wall system made out of silicon glass foam sandwiched between 22-gauge aluminum skins, and interior walls made out of aluminum honeycomb panels.[13] However, considering that many existing materials and systems are being continuously improved and new products developed on a day-to-day basis, the final spatial characteristics, construction materials and building technologies of outer space habitats may yet vary significantly from those suggested in these proposals.

The communities developed in new frontiers may also vary depending on whether the colonization efforts are made by a single nation or by a group of nations, and their success could depend on the reeducation of the dwellers to enable them to accommodate themselves to new living environments. Obviously such reeducation cannot be accomplished within the span of one generation if its goal is to achieve full-time living adaptation, so the successful implementation of many of these new habitats still lies far beyond the immediate future.

There are also some basic human characteristics and limitations which would have to be modified to permit the successful development of many of these new habitats. Traditionally, frontiersmen accepted less than adequate spatial environments and hostile living conditions because of the hope of eventually attaining better life-styles, or simply because their place of origin did not offer better opportunities. Generally, the move to new frontiers has been initiated by the few dispossessed, less qualified or needy sectors of the population. However, most of the new horizons presently considered by man require precisely the opposite type of colonist: a highly qualified scientist or technologist capable of managing, operating and servicing extremely expensive and complex equipment assemblages and habitation units, under rigorous schedules and life-styles which would offer little room for carelessness or error. This is hardly the classic image of a frontiersman.

To provide these highly qualified colonists one would have to make dwellers having a higher quality life-style abandon their traditional milieus in exchange for a foreign environment. Thus the move to inhospitable frontiers by the adequate population would only be logical after an eventual decay of most habitable surroundings on Earth. Paradoxically, it is here that the optimistic long-range visionaries and the gloom and doom prophets of mankind's future finally meet.

18
Legacies of Growth

Since the dawn of history, growth has forced man to acquire a deeper understanding of the time and space frameworks in which he lives and of his own influence upon them. Originally, Earth and infinity were one in the mind of man, and the belief that no limits existed beyond the horizon—that "infinity" actually began there—was so widespread that less than five centuries ago it almost foiled Christopher Columbus's attempt to reach India westbound from Spain.

In those days, the world known to mankind was thought to contain inexhaustable resources that could support man's expansionist tendencies indefinitely. With time, however, the physical limitations of Earth have been established. A better understanding of the world and its essential features has made it much more comprehensible and thus defined its boundaries.

Growth involves progressions and regressions. Despite its traceable patterns, it does not unfold along rectilinear paths. Thus the overall process of growth has been said to exhibit continuous sequences of forward and backward development.[1]

In modern societies growth has been viewed alternatively as the preamble to doomsday or the road to happiness. Obviously, because of the complex path of progressions and regressions exhibited by growth patterns, it is relatively easy to support either viewpoint. But in most cases, the basic question to be answered is usually ignored: Do modern societies really understand growth contextually?

Rome rose to power as a result of the same growth pattern that destroyed it. As it climbed to excessive urban concentrations, its nearby resources were progressively exhausted, thus forcing the empire to expand the limits of its domains more and more to ever greater distances and remote regions. Without slavery, its massive public work projects and its imperial army, Rome could have never reached its extensive urban development. Once set in motion, however, the unstoppable growth of Rome's imperial systems became its own demise. The larger the urban population base, the more pressing the need for increased resources and supply lines and the greater the disorder, control measures, social fragmentations, and political complexity. These chaotic conditions, in fact, led to the internal collapse of an empire which had grown too big to survive. After its fall, the population of Rome decreased from one million to 30,000 inhabitants.[2]

To imperial Rome, growth was both friend and foe, progenitor and executioner. Modern society suffers from many similar characteristics. Gradually, its development has led from sparsely populated agrarian settlements to expanding rural and mercantile communities, and eventually to huge urban concentrations. In 1900, 15 percent of the world population was urban; just over half a century later, the percentage rose to 33 percent. And it is expected that by the year 2000 there will be more urban dwellers in the world than the total Earth population of 1960.[3] These massive human concentrations, on the other hand, do not distribute themselves proportionally. As much as 80 percent of the American population lives in urban areas and more than one-half of the people of this nation live on just 1 percent of the land.[4]

The effects of these disproportionate growth patterns are serious matters. On the average, a city of a million people consumes 9,500 tons of fuel and 625,000 tons of fresh water daily. American buildings consume 57 percent of all the nation's electricity (almost one-half of this amount accounted for by lighting alone).[5] Modern edifices, in fact, require artificial lighting while occupied no matter what the light levels are outside. The progressive growth of average floor sizes has also aggravated these matters significantly: in just a few years the average office building's typical floor area has grown from 8,000–12,000 square feet to 25,000 square feet, and it is believed that shortly the figure may reach 50,000 or even 100,000 square feet.[6]

One of the most fundamental problems regarding expansions of this nature is that in modern society there seems to be a close relationship between status and growth, either due to the legitimate need for functional multiplicity and development or simply because of the necessity to keep up with surrounding growth. Thus in most modern operations the growth concept is automatically built in. This attitude is also common in speculations about future habitats. All the proponents of new and revolutionary human dwelling concepts, such as biocities, underground habitats and mega-structures, base their proposals on the assumption of built-in growth patterns very similar to those that caused the original problems the proposed "solutions" were intended to resolve. Consequently, in the majority of these proposals the critical transition processes that modify individual habitation characteristics and progressively define small colonies, cities and urban conglomerates are completely ignored.

In many cases, attempts to correct dislocated growth syndromes have been made by trying to "order" growth within carefully planned development programs. In general, these attempts are based on concepts that propose controlled evolutionary patterns or on others which presume limits and basic guidelines to restrict growth. However, this brings up another question: Can evolutionary growth be logically planned or controlled by man? There are no indications whatsoever that this is even remotely possible. Aside from the fact that it has rarely been attempted, history indicates that evolutionary patterns have never been willfully controlled by societies. It is relatively easy to illustrate the devastating effects of disproportionate growth on the Roman Empire, but to point out where the courses could have been corrected is another matter. Ultimately, to maintain that the fall of Rome could have been avoided is mere speculation.

Growth, however, does not simply happen. It is *made* to happen. Therefore, it must be possible to plan for it. But before attempting to do so, one must accept an essential premise: subordinating evolutionary patterns to conceptual frameworks means sailing into uncharted waters. Thus extreme caution must be exercised.

One of the most fundamental problems one is confronted with when trying to understand growth is that, by definition, the concept has two accepted meanings: one is the gradual *development towards* an objective, and the second is a *physical increase* not necessarily aimed at a goal. Obviously, defining growth in terms of its social implications is a complicated task from the start since, conceivably, the second interpretation of growth could even be considered opposite to the first. Thus, in the end, one is left with two basic types of growth: growth as *development* and growth as *multiplicity*.

Because of this, the relative adequacy or inadequacy of future growth patterns may vary depending on specific interpretations. For example, many forecasts look upon growth as an essential characteristic of future developments. In general, these forecasts consider the unfolding of trends and events in a context of organized sequences. Other less optimistic forecasts, on the other hand, look upon growth as an uncontrollable and erratic multiplication of items and issues that could eventually engulf everything in its path. Both overviews could be realistic and both could present *logical* scenarios of the effects of growth patterns upon society. Each of these views, however, is essentially talking about a different thing.

Bigness and smallness have been the central issues of many recent debates dealing with the effects of growth, but the fact is that the acceptance of the one does not necessarily eliminate the validity of the other and vice versa. Most small restaurants and shops that offer the "personalized" attention so highly valued in modern cities would never be feasible were it not for the large population base of contemporary urban conglomerates. The birth and expansion of suburbia was essentially the placement of decentralized smallness near centralized bigness. Satellite towns and communities, in fact, depend entirely on their nearness to densely populated urban nuclei. This is growth with purpose. Theoretically, if excess urban growth could continue to decentralize indefinitely into satellite communities, many of the maladies presently afflicting American cities could eventually be brought under control. The problem is that, to date, most urban growth problems have been caused by an increasing centralization of operations over a continuously increasing population base. Seen in perspective, the "decentralization" of functions of most urban conglomerates has been minor and merely concentrated on habitation and its supporting facilities, with little change in the primary functions that cause the concentration of population within fixed spatial parameters: occupational patterns. In this sense, modern man has differed little from the factory worker of the Industrial Revolution. Present indications, however, seem to point to increased occupational decentralization trends, and this could help solve many of the problems presently affecting urban conglomerates.

But with regard to human built environments decentralization alone is never enough to bring growth patterns under control. Social break-downs are highly accentuated by regional sectorizations and thus become dangerous corollaries of excessive growth. Most large urban nuclei, for example, are already so fragmented that their actual bigness can be defined as a conglomerate of small groupings whose decentralization could end up doing more harm than good. The fact is that beyond certain limits, urban growth (whether centralized or decentralized) becomes a mere multiplication of problems.

Relative to built environments, growth and change are also two very distinct issues, and the development of one does not necessarily keep pace with the other because, although any amount of growth inherently presumes a certain rate of change, in built environments change is not as sudden as it often is in other areas of human endeavor. Technological development, increased demand and improved production techniques, for example, have resulted in mushrooming developments in the textile and garment industries and reached a point where modern clothing is discarded long before it has worn out simply because of stylistic changes, in contrast to medieval times when articles of clothing were so expensive that they were passed on from generation to generation. If urban dwellers (probably the people least susceptible to spatial and environmental change) had to face completely new surroundings every year, they would most likely lose the few points of reference that help them situate themselves within specific time and space frameworks, and would gradually become more disoriented as the settings were progressively changed around them.

The progressively faster transformation of urban settings may end up doing just that. In many ways the rate of progress of modern societies is measured by their capacity to make their inheritance obsolete. Ancient civilizations took centuries to build, but the industrialization of underdeveloped nations can be accomplished in a few decades.[7] In this sense, modern generations have had to adapt themselves to rates of spatial change that far exceed anything previously experienced by mankind, change not only in that which is newly erected but also in that which is forever razed.

In general terms, the essential elements of the growth patterns observed in human built environments may be defined as base, demand and objectives. For example, a base consisting of plentiful resources and growing technologies propelled by the demands of increasing populations and territorial expansionism, together with the pursuit of expanding economies, scientific advances and new horizons, caused the radical changes in shelters, life-styles and urban development during the Industrial Revolution.

In this respect, adequate growth patterns can be defined in the same general terms as balanced systems: sets of related parts with common characteristics or functions that operate jointly within predefined boundaries. This is probably one of the most complex issues involved in the concept of comprehensive growth in built environments. In an attempt to balance complex, spatial characteristics and requirements, for example, growing urban nuclei have traditionally tried to define boundaries by incorporating green belts, building type groupings, and the like, with little success. Fundamentally, the problem has been one of erratic centralizing and decentralizing processes acting simultaneously upon urban fabrics.

The process of growth can affect change in built environments by altering the relationships of their components in three separate areas: proportions, apportions and levels. When cities reach sizes where the proportion of expressways and other rapid communication arteries requires cyclical intervals of repair and maintenance operations that end up invalidating the use of at least one expressway at all times, the effectiveness of massive public works projects becomes impaired by size, since the breakdown process must then be built into every new expansion.

Apportions, on the other hand, can modify growth patterns significantly. A 12 bedroom, 8 bath residence is *not* four times more expensive to build and maintain than a 3 bedroom, 2 bath house, but much costlier than that. Insurance costs, tax structures, levels of quality, maintenance and related amenities would actually cause an inbalance that could easily exceed an eight or ten multiplication factor. Thus while the proportional increases relative to landscaping, energy consumption, or some other easily quantifiable items could be defined in terms of occupancy by applying simple multipliers, the apportional factors involved in maintenance, repairs, initial costs, security and depreciation cannot be so easily defined.

Levels also characterize and define the adequacy of growth processes. What is good at the local level may be disastrous on a national scale; what is adequate for the city may be inadequate for the town, etc.

The consequences of the processes of growth in human built environments can be expressed in degrees of complexity, fragmentation, specialization, controls, multiplicities and built-in growth or change processes observed in the systems so affected.

The degree of complexity caused by growth in human built environments is accentuated by three basic principles.[8] First, most edifices must bridge a gap to increasingly refined and complex systems while relying on systems at other levels (utilities, transportation, existing structures) whose development or characteristics may not facilitate this integration. On Monday, February 28, 1983, the last episode of the television show "M.A.S.H." was aired in the United States. That day, more viewers watched that show than any other program in the history of American television. As a result, a curious phenomenon was observed in New York City: the largest demand ever recorded on the city's water supply system took place. Officials concluded, after comparing times and schedules, that the upsurge coincided with the television show ending and was due to the simultaneous use of bathroom facilities by New Yorkers at the end of the program.

The second principle is the demand made on most modern edifices to serve progressively more complex functions and to incorporate within their parameters an increasingly larger number of ancillary areas and subfunctions. Thus not only have modern habitats become increasingly dependent on complex systems, but they have also become complex systems themselves. In log cabins, for example, logs were logs that were merely placed in a convenient arrangement to provide shelter; but *logs*, nevertheless, whose modified use could take many forms. A modern sheet of metal siding, however, is *nothing* and can be made into nothing if its unique function of substitute for wood siding boards is eliminated.

Third, there are the economic factors that reflect the effects of the former principles on edifices and that, in turn, end up defining new characteristics in human habitats as well. Precisely because of changing conditions in tax structures, life cycle costs, and cost effectiveness of construction systems and materials, the evolution of many building types and urban sectors has been drastically redefined in recent years.

Complexity in built environments inevitably has a negative impact on fragmentation, specialization, controls and regulations, and built-in spatial growth and change conditions by increasing their effects upon urban fabrics and dislocating their interactive processes. In large cities, for example, constant upgradings of utility systems create continuous processes of repair and maintenance activities on a year round basis. New York—a city whose municipal workers tripled within the last decade—requires approximately 12 billion dollars a year for the repair and maintenance of its physical plant alone.[9] Developments of this sort affect the cost of city life significantly; according to the Urban Institute, the average resident of a city of one million will typically pay three times more in taxes than the resident of a city of 50,000 people.[10]

As far as habitats are concerned, increased fragmentation could easily dominate the American built environments of the future. The widespread use of home computers and improved communication channels may facilitate the increase of work-at-home patterns, but only in certain cases. The mobility of shelters, changing occupational characteristics and new habitation frontiers may take hold in future years, but only on a partial basis and only for specific sectors of the population.

One outstanding artifact intended to illustrate the effects of growth on a worldwide scale was developed by Buckminster Fuller about 30 years ago in what he called "geoscopes," spherical structures that allow all of Earth's data to be properly illustrated by computer-controlled projections on their surface. Models as large as 200 feet in diameter have been considered. At that scale, even individual building structures (specks of 1/100 of an inch) could be seen with the naked eye by merely projecting upon the sphere's surface mosaics of the 35mm pictures taken by the United States Air Force reconnaissance planes. These proportions would allow the projection of the entire city of Los Angeles to fit within a circle approximately 6 inches in diameter, and towns of 5,000 people within one-inch circumferences.[11] In geoscopes, computer-controlled models and projections could easily illustrate the effects and interfaces of forecast growth and change patterns at almost any desired level.

Ultimately, there is no reason why the same principle could not be applied in studying the impact of urban growth and population shifts upon land, transportation, housing and utilities on a nationwide basis. Concepts like this have, in fact, often been applied in urban and site planning studies, but always in rough form and without the capability of mapping all the interactive factors that the use of computers would permit. With the potential of electronics, the future of this type of study could be very promising.

The growth patterns of human habitats possess some unique characteristics that distinguish them from other areas of human endeavor. Since human scale is an overwhelming determining factor in the development of edifices, excessive growth in man's built environments sometimes ends up obstructing the basic objectives behind shelters' development. The New York 1977 blackout was the result of precisely this type of condition. Eleven computer-controlled generators linked by complex back-up systems and fail-safe provisions collapsed like lined-up dominoes because of the damage caused to some power lines by a thunderstorm and a couple of minor malfunctions. One serious point relative to the complexity of overgrown systems of this nature is that in many cases they are not designed *ever* to stop operating, so that total system failure is never programmed or considered. In New York's case, there had not been any provision made for the start-up of the total system should it ever fail.[12]

Urban utilities and services, however, are mere support functions for the development of edifices. They are a direct consequence of the explosive sizes of modern building projects and are ultimately tied to their unstoppable growth. The Sears Tower in Chicago, for example, consumes more electricity than the entire city of Rockford, Illinois (147,000 inhabitants),[13] and the length of its elevator cables alone could reach Milwaukee. Yearly elevator travels throughout the United States, in fact, average about 10,000 miles per year, the distance between New York and Australia.[14]

Significant facts and figures relative to the disregard for the human scale in modern buildings and urban settings have been documented in many studies. Over 20 years ago, analyses made by H. Dreyfus found that the optimum vertical viewing angle of the human eye was between 25 and 37 degrees, thus limiting the overall size of a building to two stories (approximately 25 feet in height) if one is to see the full elevation at a horizontal distance of 50 to 55 feet (an average 4-lane suburban street/sidewalk combination). Also the horizontal viewing angle was determined at 60°, thus limiting the width of the building from 50 to 60 feet at the distance of 50 to 55 feet from the observer.[15]

Previous studies had also proven that a 72-foot distance was the maximum permissible for recognizance of human features and that 48 feet was the maximum allowable distance for appreciating facial expression. Thus if street widths were determined on these bases (and also employing visual angles defined by Dreyfus), 72-foot-wide streets would limit the height of structures to 30 feet.[16]

Four hundred and fifty feet is considered the maximum distance allowable for the human eye to distinguish colors, approximate age and clothing of a person. Distances that range anywhere from 430 to 465 feet have also served as ideal measurements for most public spaces and plazas throughout history. Venice's Plaza St. Marcus is 425 feet across, and St. Peter's in Rome is 430 feet wide. Thus if an average 450-foot measure is subjected to the same angle of view analysis, the maximum height for buildings in open public spaces turns out to be 225 feet (18 stories more or less),[17] which is a far cry from the average observer-to-building-height proportions of modern urban conglomerates.

Added to these physical characteristics are the kinesthetic factors. Multiple sets of equal elements and modules endlessly repeated throughout modern building projects and can only be appreciated by accelerating the natural motion through space of human beings, thus altering the natural perception of human kinetic movement in and around edifices with imagery grasped at "automotive" speeds. It would take anyone almost three times longer to walk the perimeter of the Ford Motor Company Parts Redistribution Center in Michigan than to stroll around the pyramid of Cheops in Egypt, whose area, at the base, would fit well over five times inside the perimeter of the former.[18]

The fact is that the sizes of many modern edifices are well beyond the limits suggested by human scale. One could fit over 3,600 2-story, 4 bedroom houses inside the John F. Kennedy Space Center Vehicle Assembly Building in Florida.[19]

The continued growth of many building types has also affected the scope of most construction and real estate investments which have become the primary forces behind most urban and suburban land development projects. The actual economic life of a shopping center (and other similar groupings of commercial/retail centers) is presently considered to be under 20 years. This is due to changing consumer needs and the resulting increases in the average sizes of most retail establishments during the past few decades: grocery stores from 13,000 to 25,000 square feet and department stores from 75,000 to 180,000 square feet, and the end is nowhere in sight; it is already commonly assumed that the typical shopping mall of the 80's will rarely cover less than 1,000,000 square feet of floor area.[20]

In modern society there are several key factors that limit the relative size of buildings. Some of these factors and the way they impose spatial limits upon habitats are as follows:

1. Regulatory issues: distances to exits, parking and zoning requirements, limits to floor areas and number of stories, etc.

2. Feasibility matters: cost/time recuperation factors, financing, time of construction, etc.

3. Local confinement: peripheral limits, nonexpansive conditions, etc.

4. Related factors: transportation arteries, utilities and services, maintenance and operational costs, security, etc.

However, no matter how drastically some of these factors may influence the size of edifices at one time or another, traditionally they have always given way to economic pressures, new technologies and operational changes, and have progressively allowed for spatial increases.

The fact is that in modern shelters growth no longer depends on physical limitations but merely on the operations that those structures envelop. Because of this, certain building types tend to fragment into smaller units once they reach a specific size, as in the case of banks, gas stations, and restaurants (not surprisingly, those building types that currently have the lowest square foot averages throughout the United States)[21] while others, such as manufacturing plants, entertainment centers, and shopping malls have followed the opposite course.

It is because of this basic dependency of the size of shelters on increasingly complex operational developments that the traditional growth-limiting factors cited above consistently concede spatial increases to almost all building types. Accordingly, one finds that as long as edifices conform to pre-established fire ratings or rules governing distance to exits and are located within adequate fire zones, there is always a building code section that will read "unlimited" on the tables that define the allowable size of buildings, thus accepting that conceptually there are no parameters beyond which edifices *cannot* grow. Such is not the case however; buildings—just like cities and communities—*do* have limits and levels of efficiency depending on operational and functional adequacy beyond which their ultimate performance begins to decline.

One of the most complex goals relative to the growth patterns of building types is the achievement of an adequate balance between function and size, because in edifices, "back to small" impositions or arbitrary limits to growth concepts are quite dangerous and, in most instances, simply will not work. In fact, most new buildings developed in strict accordance with the human scale figures previously mentioned would be totally unfeasible 9 out of 10 times due to economic factors, land availability constraints, and basic built-in growth requirements. Ultimately, the combination of these issues makes reduced scale buildings more the exception than the rule. Winston Churchill was quoted as saying: ''We shape our buildings and thereafter our buildings shape us.''[22] In many ways modern man is imprisoned in a spatial vicious circle where edifices denounce the inadequacies of the past, but do so without regard for what may become tomorrow's guidelines. Undeniably, the conception and development of modern built environments has become essentially reactive.

The ultimate effects of growth on human built environments, however, do not always present themselves as out-of-proportion developments where overwhelming bigness is a basic characteristic. In many instances their most significant consequences are strictly due to erratic evolutionary trends rather than overwhelming multiplicities.

From the outset, urban growth—no matter how small—was the driving force behind the decentralization of work centers that eventually made the majority of mass transit systems unfeasible in large cities. With private automobiles as the only viable means of rapid communication and urban work centers becoming increasingly decentralized, the process has consistently been accelerated. The automobile was introduced as a means of reducing travel times, but the urban and suburban decentralization patterns triggered throughout America by the widespread use of private cars have achieved exactly the opposite. The one-half hour to 45 minutes that most urbanites spend today in travel to and from work centers is no different from the time it took workers to commute to work 40 years ago by mass transit systems.[23] The cost burdens imposed upon the American consumer by this change, however, have been staggering. On the average, car owners consume 25 percent of their income for automobiles (more than for food). Between the mid–50's and 1970 the United States spent almost 200 billion dollars on highway construction, and conservative estimates show that expenditures of up to 300 billion dollars may be allocated between 1973 and 1985 for highway related work.[24] The heavy gasoline tax recently proposed to create jobs in a slumping economy, for example, will be solely dedicated to highway extension, repair or maintenance work—no small task, by any means: the United States interstate highways already extend 42,500 miles (9 times the distance between New York City and Moscow). With 3,600,000 square miles of land, the United States has over 3,600,000 miles of roadways.[25] The American roadway system alone could circle the Earth almost 150 times!

Manufacturing, maintenance and storage of motorized vehicles has soared to levels almost beyond control. Automobiles alone consume 20 percent of all the steel, 51 percent of the lead, 95 percent of the nickel and 60 percent of the rubber used in the United States; and on the average, every day 10,000 new cars are added to the American roadway system.[26] That in itself represents an average of almost 4 million square feet taken up by new automobile storage *daily*.

Modern construction projects are commonly criticized for sacrificing natural features and landscapes to parking lots. However, parking lots are only built because modern life-styles *demand* them. A shopping mall with beautifully landscaped gardens and inadequate parking would most likely go out of business during its first year of operation; so would any hospital, apartment complex or office building that lacked parking space. And it is very unlikely that any one of these projects could ever find financing or be granted a building permit if it did not provide adequate parking facilities. The fact is that for most edifices, parking lots are merely inconvenient, costly and undesirable satellite services that *must* be provided simply because of the demands of modern life-styles.

In context, the growth of human built environments has led to accelerated fragmentation and distribution patterns, and caused significant territorial modifications. And these, in turn, have affected the economies, resources and the technoscientific progress that ultimately define the characteristics of habitats. Thus, in a broad sense, the repercussions of the growth patterns of habitats finally end up redefining their own development matrix.

Since the colonial period, for example, continued economic growth has increased the fragmentation of this nation's economic processes, but it has also improved the distribution of wealth. By doing so, it has reduced socioeconomic differentiation, but it has accentuated territorial and occupational differences.

The demands imposed upon the exploitation of resources by continued economic development and population growth, on the other hand, have eventually contributed to facilitate resources availability. Unfortunately, this has been done sometimes at the expense of fixed or nonrenewable bases. Growing demands have further underlined regional characteristics and territorial specializations.

Technoscientific growth has increased occupational fragmentation, contributed to a better distribution of wealth and goods, and reduced cultural fragmentation. It has also contributed to a decrease of territorial differences and indigenous characteristics.

Thus, comparing the effects of these interactive issues and the probable trends of the major factors (a relative decline in economic development and resources availability and continued technoscientific growth), one can conclude that the general effects of growth patterns most likely to affect the evolution of habitats in future decades could involve further socioeconomic fragmentation and cultural unification; slight declines in resources availability and increased exploitation processes; rises and declines in regional economies; and a reduction of territorial influences on built environments.

Another concern regarding the adequacy of tomorrow's shelters is the rate at which future evolutions may occur. Never before has time been such an increasingly significant factor. In modern society, many spatial "futures" are being built at an overwhelming rate, but one must inevitably face the question of their ultimate significance on the evolutionary ladder: do they really represent spatial "futures" or are they merely wishful of unrealistic preconceptions rushed upon everyone?

Novelty, for example, suggests (and may even produce) full "video screen walls" in future shelters. Many forecasters consider this a likely possibility.[27] However, this could be exactly the kind of wishful prediction made without a careful analysis of the implications involved, an example of the direct application of growth concepts without the full knowledge of facts.

By all present standards and limits imposed by human scale, full video screen walls would require unobstructed distances of no less than 10 feet (assuming an average 8 by 14 foot living room wall). This would presume a total spatial requirement of at least 140 square feet in the average dwelling unit dedicated to television viewing *only*. Not only an extremely unrealistic assumption, but one that would require the average viewer to sit inside a 2-foot deep pit, since to watch any large screen comfortably the observer's eyes should be at approximately the same level as the bottom of the projection. Conversely, if one uses the present average viewing distance of 7 to 8 feet from the television set, the maximum viewing size of any TV screen would be approximately 3.5 to 4 feet high and slightly curved to compensate for glare and distortion, a shape and dimensions which, curiously enough, closely coincide with portable and convenient giant screen television sets presently available.

When it is forecast that housing units could become full-time work centers penetrated by a myriad of communication channels and external intrusions, or that the concept of "home" could be obliterated by scheduled space/function habitation patterns, and such developments are looked upon as the logical outcome of present trends, the future of habitats illustrates much more than the mere correspondence between resources availability, socioeconomic conditions and technoscientific achievements.

Future scenarios of electronic cottages, work/live-in complexes, underground cities and outer space colonies may resolve many current problems, but they could also demand radical changes of questionable value in human nature. If the basis for human habitats is transformed from the development of multipurpose living environments to the mere supply of spaces intended only to satisfy rigidly scheduled operational functions and habitation patterns, man himself will have changed radically since everything he has ever believed in, worked towards or achieved in his shelters will have been cast aside, making his built environment a mere function of operational sequences. Thus edifices would finally become integral parts of scientifically conceived and generated systems. Shelters, in fact, would become part of man's machines. Then, with no place left to retreat, the man of the future might realize that the only escape left from time was time itself, since in such a world there would be no more *places* to hide, only *times* to hide.

Maybe it is precisely because of the scientific miracles presently unfolding before modern society that the future appears to hold more dangers and uncertainties than it ever did before. When one can anticipate such wonders as outer space colonization and automated environments, but must also consider the possibility of nuclear holocausts and world famine as logical developments of present trends, an urgent necessity to seriously reflect upon heritages, evolutions and objectives becomes apparent.

More than a time for action, it may be a time for thought.

Notes

INTRODUCTION

1. Gerard O'Neill, "Near-Future Impact on Rapid Technological Change," 1982 AIA National Convention, Honolulu, Hawaii.
2. Alan P. Lightman, "Cosmic Natural Selection," *Science '83* (July/August 1983): 26.

1. THE ORDER OF DISORDER

1. Frank Mankiewicz and Joel Swerdlow, *Remote Control*, p. 68.
2. Bernard J. Frieden, *The Environmental Protection Hustle*, p. 12.
3. "First-Time Market," *Professional Builder* (January 1981): 208, 216.
4. Dean S. Rugg, *Spatial Foundations of Urbanism* (Dubuque, Iowa: Wm. C. Brown Company Pub., 1979), p. 197.
5. "How Mortgage Money Gets Where It's Needed," *Professional Builder* (March 1981): 41.

2. FACTS ON FILE

1. Robert Stobaugh and Daniel Yergin (eds.), *Energy Future*, p. 207.
2. Alfred V. Zamm, *Why Your House May Endanger Your Health*, pp. 29–43.

3. Ibid., pp. 61–87.

4. Robert C. Cowen, "U.S. Faces Shortage in Critical Raw Materials," *Wisconsin State Journal* (March 1981).

5. Zigurds Grigalis, "Material Resources," *The Construction Specifier* (June 1980): 52.

6. Based on data provided by Heron House Associates (eds.), *The Book of Numbers* (New York: A & W Publishers, Inc., 1978), pp. 158–59.

7. Kirkpatrick Sale, *Human Scale*, p. 109.

8. Saul Pett, "Uncle Sam: The Incredible Bulk," *Arizona Republic* (June 14, 1981), p. AA1.

9. Ibid.

3. FORMS AND FUNCTIONS

1. Glenn H. Beyer, *Housing and Society*, p. 5.

2. Ibid., p. 19.

3. James Deetz, *In Small Things Forgotten*, p. 1.

4. Beyer, *Housing and Society*, p. 39.

5. Ibid., p. 39.

6. Ibid., p. 34.

7. Ibid., p. 35.

8. Ibid., p. 35.

9. Marcus Whiffen and Frederick Koeper, *American Architecture 1607–1976* (Cambridge: The MIT Press, 1981), p. 238.

10. Beyer, *Housing and Society*, p. 38.

11. Richard Smith, et al., *The Average Book* (New York: Rutledge Press, 1981), p. 220.

4. GROUPINGS: DYNAMICS AND PROFILES

1. Arthur B. Gallion and Simon Eisner, *The Urban Pattern*, p. 12.

2. Ibid., p. 30.

3. Ibid., p. 31.

4. Ibid., p. 36.

5. Glenn H. Beyer, *Housing and Society*, p. 448.

6. J. Franklin Jameson (ed.), *Narratives of New Netherland 1609–1664* (New York, 1909), pp. 111–12.

7. Gallion and Eisner, *The Urban Pattern*, pp. 49–52.

8. Beyer, *Housing and Society*, p. 39.

9. Frederick Lewis Allen, *The Big Change*, p. 19.

10. Gallion and Eisner, *The Urban Pattern*, pp. 81, 82.

11. Saul B. Cohen (ed.), *Problems and Trends in American Geography* (New York: Basic Books, Inc., Publishers, 1967), p. 48.

12. Edward T. Hall, *The Hidden Dimension*, p. 178.

13. Ibid., p. 106.

14. Gallion and Eisner, *The Urban Pattern*, p. 56.

15. Hall, *The Hidden Dimension*, p. 175.

5. THE PROCESS OF MODERNIZATION

1. Charles H. Hession and Hyman Sardy, *Ascent to Affluence*, p. 87.

2. W. W. Rostow, *The World Economy*, p. 142.

3. Hession and Sardy, *Ascent to Affluence*, p. 32.

4. Glenn H. Beyer, *Housing and Society*, p. 36.

5. Thomas C. Cochran, *Business in American Life* (New York: McGraw-Hill Book Company, 1972), p. 67.

6. Ibid., p. 67.

7. Hession and Sardy, *Ascent to Affluence*, p. 175.

8. Ibid., pp. 520–21.

9. Rostow, *The World Economy*, pp. 243–44.

10. Lester C. Thurow, *The Zero-Sum Society*, pp. 48, 52.

6. THE HUMAN CONDITION

1. Hannah Arendt, *The Human Condition* (Chicago: The University of Chicago Press, 1958), p. 30.

2. Ibid., p. 32.

3. Kirkpatrick Sale, *Human Scale*, p. 165.

4. Glenn H. Beyer, *Housing and Society*, p. 52.

5. Ibid., p. 52.

6. Ibid., p. 68.

7. Ibid., p. 77.

8. Heron House Associates (eds.), *The Book of Numbers* (New York: A & W Publishers, Inc., 1978), p. 142.

9. Gerald M. McCue, et al., *Creating the Human Environment*, p. 61.

10. Jerry Mander, *Four Arguments for the Elimination of Television* (New York: Morrow Quill Paperbacks, 1978), pp. 61–65.

7. SOCIAL STATUS AND STRATA

1. Richard P. Coleman and Lee Rainwater, *Social Standing in America*, p. 7.

2. Ibid., p. 296.

3. Ibid., pp. 301–3.

8. POLICIES, POLITICS, AND PROPAGANDA

1. John Bartlett, *Familiar Quotations*, Emily Morison Beck (ed.) (Boston: Little, Brown and Company, 1980), p. 136.

2. Glenn H. Beyer, *Housing and Society*, p. 449.

3. Ibid., p. 450.

4. Ibid., p. 452.

5. Ibid., p. 454.

6. Ibid., p. 455.

7. Construction Sciences Research Foundation, *Construction Trends and Problems Through 1990*, p. 12.

9. BUILDINGS AND CONSTRUCTION

1. William J. Mitchell, *Computer-Aided Architectural Design* (New York: Petrocelli/Charter, 1977), pp. 73–74.

2. Heron House Associates (eds.), *The Book of Numbers* (New York: A & W Publishers, Inc., 1978), p. 39.

3. Ibid., pp. 23, 145.

4. Gerald M. McCue, et al., *Creating the Human Environment*, p. 191.

5. Ibid., p. 205.

6. Ibid., p. 205.

7. David Reuben Michelsohn, et al., *Housing in Tomorrow's World*, p. 13.

8. Ibid., p. 25.

9. Caleb Hornbostel, *Construction Materials* (New York: John Wiley & Sons, 1978), p. 246.

10. Glenn H. Beyer, *Housing and Society*, p. 203.

11. Ibid., p. 228.

12. Elena Zucker (ed.), "CII Loans," *Building Design and Construction* (January 1980): 56.

13. Daniel J. Boorstin, *The Republic of Technology*, p. 9.

14. Richard Smith, et al., *The Average Book* (New York: Rutledge Press, 1981), p. 172.

15. Arthur B. Gallion and Simon Eisner, *The Urban Pattern*, p. 257.

16. Ibid., p. 258.

17. Ibid., p. 272.

18. Heron House, *The Book of Numbers*, p. 197.

19. U.S. Department of Commerce, Bureau of the Census, *1980 Census of Population and Housing* (April 1981), p. 4.

20. Stephen Rosen, *Future Facts*, pp. 235–37.

21. Ibid., pp. 235–37.

10. TIME AND TIME MACHINES

1. Edward T. Hall, *The Hidden Dimension*, p. 173.

2. Joost A. M. Meerloo, *Along the Fourth Dimension*, p. 125.

3. Ibid., pp. 27, 134.

4. Ibid., p. 97.

11. FUTURISM AND FORECASTING

1. "Acid Rain Debate Gears Up," *Wisconsin State Journal* (April 11, 1981).

2. Edward S. Cornish, "Preview of the Next Millennium," Session #1, Neocon 12, Chicago, 1979.

3. Robert Stobaugh and Daniel Yergin (eds.), *Energy Future*, p. 305.

4. Cornish, "Preview of the Next Millennium."

5. Stobaugh and Yergin, *Energy Future*, p. 305.

6. Gerald M. McCue, et al., *Creating the Human Environment*, p. 141.

7. Based on Robert U. Ayres, *Uncertain Futures*, p. 2.

8. Based on Cornish, "Preview of the Next Millennium."

9. Russell V. Keune, "Existing Building Stock," 1982 AIA Conference, Honolulu, Hawaii.

12. ECONOMY, RESOURCES, AND TECHNOLOGY

1. Richard Smith, et al., *The Average Book* (New York: Rutledge Press, 1981), p. 165.

2. John A. Loraine, *Global Signposts to the 21st Century*, p. 169.

3. W. W. Rostow, *The World Economy*, pp. 614–15.

4. Smith, et al., *The Average Book*, p. 67.

5. Bruce K. Ferguson, "Whiter Water," *The Futurist* (April 1983): 30–31.

6. Smith, et al., *The Average Book*, p. 63.

7. Ibid., p. 217.

8. Ibid., p. 165.

9. Robert A. Lemire, *Creative Land Development*, pp. 14–16.

10. Smith, et al., *The Average Book*, p. 172.

11. Lemire, *Creative Land Development*, p. 16.

12. William K. Hartmann, "Mines in the Sky," *Smithsonian* 13 (September 1982): 73–74.

13. SOCIOLOGICAL ISSUES

1. Robert Reinhold, "Electronically Run Future Envisioned," *Arizona Daily Star* (Tucson) (July 4, 1982).
2. Koichi Ura, "Building Markets Overseas," *The Construction Specifier* (February 1981): 17.
3. Richard Smith, et al., *The Average Book* (New York: Rutledge Press, 1981), p. 59.
4. Alvin Toffler, *The Third Wave*, p. 201.

14. GUIDELINES

1. U.S. Department of Commerce, Bureau of the Census, *U.S. Department of Commerce News*, (Washington, D.C.: February 23, 1981), p. 1.
2. Richard Smith, et al., *The Average Book* (New York: Rutledge Press, 1981), pp. 60, 63.
3. "How Mortgage Money Gets Where It's Needed," *Professional Builder* (March 1981): 41.
4. Smith, et al., *The Average Book*, p. 56.

15. HABITATS: NEW HORIZONS

1. "Dramatic Changes Coming for Bathroom Interiors," *The Futurist* (April 1976): 111.
2. Harlan Cleveland, "After Affluence, What?" *The Futurist* (October 1977): 279.
3. "Snugger Harbor," *Next* (July/August 1980): 67.
4. "New Things," *Next* (March/April 1981): 8.
5. "A House That Cleans Itself," *The Gallery Gazette* (May 1982): 1.
6. David Wallechinsky, et al., *The Book of Predictions* (New York: William Morrow & Company, Inc., 1981), p. 468.
7. David Konigsburg, "Commune," *Next* (November/December 1980): 56.
8. John Reilly, in *USA Today* (August 3, 1983): 18.
9. " . . . Tomorrow's Houses," *Wisconsin State Journal* (February 11, 1980).
10. "Housing," *Daily Reporter* (Tucson) (March 3, 1982): 17.
11. "The Marketplace," *Psychology Today* (December 1982): 16.
12. David Reuben Michelsohn, et al., *Housing in Tomorrow's World*, p. 49.

13. Stephen Rosen, *Future Facts*, p. 416.

14. Roy Mason, "Underground Architecture," *The Futurist* (February 1976): 19.

15. "Construction Goes Underground," *The Futurist* (December 1979): 487.

16. Rosen, *Future Facts*, pp. 269–72; and "Subterrene Would Tunnel Deep into Earth's Crust," *The Futurist* (February 1976): 24.

17. "Technology," *Next* (November/December 1980): 100.

18. Roy Mason, "Biological Architecture," *The Futurist* (June 1977): 140–48.

19. Wolf Hilbertz, "Building Environments That Grow," *The Futurist* (June 1977): 148–49.

20. Ibid., pp. 148, 149.

21. Roy E. Mason, "Habitat 2000," *The Construction Specifier* (January 1980): 79–80.

22. Ibid., pp. 79–80.

23. Kirby L. Estes, "The Extra Dimension," *The Futurist* (December 1979): 439–44.

24. Glen Small, "Land in the Sky," *The Futurist* (June 1977): 150–53.

25. "World Trends and Forecasts: Architecture," *The Futurist* (October 1976): 283–84.

26. Estes, "The Extra Dimension," p. 444.

27. Patricia G. Tucker, "2001 (Revisited)," *The Construction Specifier* (January 1980): 122.

28. Magoroh Maruyama, "Designing a Space Community," *The Futurist* (October 1976): 279.

29. Robert M. Salter, "Planetran," *The Construction Specifier* (January 1980): 100.

30. Ibid., p. 101.

16. SYSTEMS AND PRODUCTS

1. "A Tent Bigger than the Pentagon," *Next* (March/April 1980): 111.

2. Stephen Rosen, *Future Facts*, p. 288.

3. "Getting Around," *Next* (January/February 1981): 122.

4. "Living," *Next* (July/August 1981): 84.

5. Wolf Hilbertz, "Building Environments that Grow," *The Futurist* (June 1977): 148.

6. Roy E. Mason, "Habitat 2000," *The Construction Specifier* (January 1980): 75.

7. David Reuben Michelsohn, et al., *Housing in Tomorrow's World*, p. 60.

8. Alice Koller, "Fiber Reinforced Concrete," *The Construction Specifier* (December 1982): 44–55.

9. "Living," *Next* (March/April 1981): 121.

10. William E. Sharpe, "The Future Is Water Efficient Plumbing," *The Construction Specifier* (June 1982): 26–33.

11. "Your Move," *Next* (May/June 1980): 14.

12. Richard Smith, et al., *The Average Book* (New York: Rutledge Press, 1981), p. 59.

13. "Environmental Design," *Design & Environment* (Spring 1970): 9.

14. Michelsohn, et al., *Housing in Tomorrow's World*, p. 54.

15. "Working," *Next* (March/April 1981): 110.

16. "Technology," *Next* (September/October 1980): 113.

17. Patricia G. Tucker, "2001 (Revisited)," *The Construction Specifier* (January 1980): 130.

18. "Technology," *Next* (May/June 1980): 81–82.

19. Rosen, *Future Facts*, p. 283.

20. Ibid., p. 275.

21. Smith, et al., *The Average Book*, p. 173.

22. Mark Diamond, "Teaching Elevators to Talk," *The Construction Specifier* (June 1982): 60–64.

23. Rosen, *Future Facts*, pp. 107–8.

24. Michelsohn, et al., *Housing in Tomorrow's World*, p. 31.

25. Ibid., p. 32.

26. David Lampe, "Why Not Grow a Building Underwater!" *Next* (March/April 1980): 57–62; and Wolf H. Hilbertz, "Oceantecture," *The Construction Specifier* (January 1980): 148–49.

27. Hilbertz, "Building Environments that Grow," pp. 148–49.

28. Advertisement in *The Futurist* (August 1982): 77.

29. "Society," *Next* (September/October 1980): 102.

30. "Tomorrow in Brief," *The Futurist* (June 1983): 5.

17. ORGANIZATIONS

1. Hollis Vail, "The Automated Office," *The Futurist* (April 1978): 77.

2. William L. Rentro, "Second Thoughts on Moving the Office Home," *The Futurist* (June 1982): 43–48.

3. "Architecture," *The Futurist* (April 1976): 112.

4. Based on Magoroh Maruyama, "Designing a Space Community," *The Futurist* (October 1976): 277.

5. Arthur B. Gallion and Simon Eisner, *The Urban Pattern*, p. 6.

6. Ibid., p. 5.

7. "Granny Flats," *The Futurist* (February 1982): 51.

8. Construction Sciences Research Foundation, *Construction Trends and Problems Through 1990*, p. 9.

9. Based on Maruyama, "Designing a Space Community," 281.
10. Patricia G. Tucker, "2001 (Revisited)," *The Construction Specifier* (June 1982): 129.
11. Maruyama, "Designing a Space Community," p. 281.
12. Ibid., p. 280.
13. Tucker, "2001 (Revisited)," p. 129.

18. LEGACIES OF GROWTH

1. Joost A. M. Meerloo, *Along the Fourth Dimension*, p. 224.
2. Jeremy Rifkin, *Entropy*, p. 151.
3. Ibid., p. 149.
4. Ibid., p. 148.
5. Ibid., p. 152.
6. Moshe Safdie, *Form and Function* (Boston: Houghton Mifflin, 1982), p. 133.
7. Daniel J. Boorstin, *The Republic of Technology*, p. 5.
8. Based on W. H. Mayall, *Principles in Design*, pp. 31–34.
9. Rifkin, *Entropy*, pp. 153, 155.
10. Ibid., p. 156.
11. R. Buckminster Fuller, *Critical Path*, pp. 161–84.
12. Kirkpatrick Sale, *Human Scale*, pp. 83–85.
13. Rifkin, *Entropy*, p. 152.
14. Richard Smith, et al., *The Average Book* (New York: Rutledge Press, 1981), p. 227.
15. Sale, *Human Scale*, pp. 170–72.
16. Ibid., p. 173.
17. Ibid., pp. 174–75.
18. Based on data provided by Diagram Group, *Comparisons* (New York: St. Martin's Press, 1980), p. 89.
19. Ibid., p. 92.
20. The American Institute of Architects, *Life Cycle Cost Analysis* (AIA, 1977), p. 2.
21. Based on data provided by Robert Snow Means Company, Inc., *Building Construction Cost Data 1983* (RSM Co., 1982), p. 392.
22. Sale, *Human Scale*, p. 165.
23. Rifkin, *Entropy*, p. 143.
24. Ibid., pp. 143–45.
25. Ibid., p. 145.
26. Ibid., p. 142.
27. David Wallechinsky, et al., *The Book of Predictions* (New York: William Morrow and Company, Inc., 1981), p. 468.

Bibliography

Allen, Frederick Lewis. *The Big Change*. New York: Perennial Library, 1952.

Ayres, Robert U. *Uncertain Futures*. New York: John Wiley & Sons, 1979.

Beyer, Glenn H. *Housing and Society*. New York: The Macmillan Company, 1965.

Boorstin, Daniel J. *The Republic of Technology*. New York: Harper & Row, Publishers, 1978.

Brown, Lester R. *Building a Sustainable Society*. New York: W. W. Norton & Company, 1981.

Cetron, Marvin and Thomas O'Toole. *Encounters with the Future*. New York: McGraw-Hill Book Company, 1982.

Coleman, Richard P. and Lee Rainwater. *Social Standing in America*. New York: Basic Books, Inc., Publishers, 1978.

Commoner, Barry. *The Closing Circle*. New York: Bantam Books, 1980.

Construction Sciences Research Foundation, Inc., The. *Construction Trends and Problems Through 1990*. Washington, D.C.: The Construction Sciences Research Foundation, Inc., 1981.

Deetz, James. *In Small Things Forgotten*. New York: Anchor Books, 1977.

Evans, Christopher. *The Micro Millennium*. New York: The Viking Press, 1979.

Frieden, Bernard J. *The Environmental Protection Hustle*. Cambridge: The MIT Press, 1979.

Fuller, R. Buckminster. *Critical Path*. New York: St. Martin's Press, 1981.

Gallion, Arthur B. and Simon Eisner. *The Urban Pattern*. 4th ed. New York: D. Van Nostrand Company, 1980.

Hall, Edward T. *The Hidden Dimension*. New York: Anchor Books, 1969.

Hawken, Paul, James Ogilvy, and Peter Schwartz. *Seven Tomorrows*. New York: Bantam Books, 1982.

Hession, Charles H. and Hyman Sardy. *Ascent to Affluence*. Boston: Allyn & Bacon, Inc., 1969.

Kahn, Herman. *The Next 200 Years*. New York: William Morrow & Company, Inc., 1976.

Kerr, Clark and Jerome M. Rosow (eds.). *Work in America*. New York: Van Nostrand Reinhold Company, 1979.

Lemire, Robert A. *Creative Land Development*. Boston: Houghton Mifflin Company, 1979.

Loraine, John A. *Global Signposts to the 21st Century*. Seattle: University of Washington Press, 1979.

Mankiewicz, Frank and Joel Swerdlow. *Remote Control*. New York: Ballantine Books, 1978.

Mayall, W. H. *Principles in Design*. New York: Van Nostrand Reinhold Company, 1979.

McCue, Gerald M., et al. *Creating the Human Environment*. Urbana: University of Illinois Press, 1970.

Meerloo, Joost A. M. *Along the Fourth Dimension*. New York: The John Day Company, 1970.

Michelsohn, David Reuben, et al. *Housing in Tomorrow's World*. New York: Julian Messner, 1973.

Rifkin, Jeremy. *Entrophy*. New York: The Viking Press, 1980.

Rosen, Stephen. *Future Facts*. New York: Simon & Schuster, 1976.

Rostow, W. W. *The World Economy*. Austin: University of Texas Press, 1978.

Sale, Kirkpatrick. *Human Scale*. New York: Coward, McCann & Geoghegan, 1980.

Smith, Wallace F. *Housing*. Berkeley: University of California Press, 1970.

Stobaugh, Robert and Daniel Yergin (eds.) *Energy Future*. New York: Ballantine Books, 1979.

Thurow, Lester C. *The Zero-Sum Society*. New York: Basic Books, Inc., Publishers, 1980.

Toffler, Alvin. *The Third Wave*. New York: William Morrow & Company, Inc., 1980.

Zamm, Alfred V. *Why Your House May Endanger Your Health*. New York: Simon & Schuster, 1980.

Index

About the Author

MANUEL MARTI JR. received his degree in Architecture from the National University of Mexico. He is also the author of *Space Operational Analysis: A Systematic Approach to Spatial Analysis and Programming.*